Fields Institute Communications

The Fields Institute for Research in Mathematical Sciences

Bifurcation Theory and Spatio-Temporal Pattern Formation

Wayne Nagata
N. Sri Namachchivaya
Editors

American Mathematical Society
Providence, Rhode Island

The Fields Institute for Research in Mathematical Sciences

2000 *Mathematics Subject Classification.* Primary 34–06, 35–06, 37–06.

Library of Congress Cataloging-in-Publication Data

Bifurcation theory and spatio-temporal pattern formation / Wayne Nagata, N. Sri Namachchivaya.
p. cm. — (Fields Institute Communications, ISSN 1069-5265 ; 49)
Includes bibliographical references.
ISBN-13: 978-0-8218-3725-2 (alk. paper)
ISBN-10: 0-8218-3725-7 (alk. paper)
1. Bifurcation theory—Congresses. 2. Dynamics—Congresses. 3. Pattern formation (Physical sciences)—Congresses. I. Nagata, Wayne. II. Namachchivaya, N. Sri (Navaratnam Sri), 1957– III. Fields Institute for Research in Mathematical Sciences.

QA380.B566 2006
515′.392—dc22 2006048362

∞ The paper used in this book is acid-free and falls within the guidelines
established to ensure permanence and durability.
This publication was prepared by the Fields Institute.
http://www.fields.utoronto.ca
Visit the AMS home page at http://www.ams.org/

10 9 8 7 6 5 4 3 2 1 11 10 09 08 07 06

Contents

Preface

Nonlinear dynamical systems and the formation of spatio-temporal patterns play an important role in current research on partial differential equations. Progress in the past quarter-century in the development of theories of equivariant bifurcations, mode interactions and deterministic chaos has made possible a better understanding of spatio-temporal patterns in a wide variety of physical and biological contexts.

This volume contains the proceedings of the Workshop on Bifurcation Theory and Spatio-Temporal Pattern Formation in Partial Differential Equations, held at The Fields Institute for Research in Mathematical Sciences, December 11 to 13, 2003. The Workshop contributed to the 2003–2004 Thematic Program on Partial Differential Equations at The Fields Institute along this important direction of current research and complemented a related Workshop on Patterns in Physics, held November 14 to 18, 2003. There were 27 presentations made, seven of which are represented by expository articles in this volume.

The Workshop and this volume are dedicated to Professor William Langford, one of the pioneering researchers in nonlinear dynamics and bifurcation theory in Canada. We recognize his fundamental contributions and leadership in bifurcation theory and pattern formation, and acknowledge his service to the Canadian mathematical community.

The Workshop was financially sponsored and hosted by The Fields Institute. The USA National Science Foundation and the Department of Mathematics and Statistics, University of Guelph provided additional financial support.

Wayne Nagata and N. Sri Namachchivaya

Fields Institute Communications
Volume **49**, 2006

Flow Invariant Subspaces for Lattice Dynamical Systems

Fernando Antoneli
Department of Applied Mathematics
University of São Paulo
São Paulo SP 05508-090, Brazil
antoneli@ime.usp.br

Ana Paula S. Dias
Departamento de Matemática Pura
Centro de Matemática da Universidade do Porto
Faculdade de Ciências
4169-007 Porto, Portugal
apdias@fc.up.pt

Martin Golubitsky
Department of Mathematics
University of Houston
Houston TX 77204-3008, USA
mg@uh.edu

Yunjiao Wang
Department of Mathematics
University of Houston
Houston TX 77204-3008, USA
yunjiao@math.uh.edu

Abstract. Stewart *et al.* have shown that flow invariant subspaces for coupled networks are equivalent to a combinatorial notion of a balanced coloring. Wang and Golubitsky have classified all balanced two colorings of planar lattices with either nearest neighbor (NN) or both nearest neighbor and next nearest neighbor coupling (NNN). This classification gives a rich set of patterns and shows the existence of many nonspatially periodic patterns in the NN case. However, all balanced two-colorings in the NNN case on the square and hexagonal lattices are spatially periodic. We survey these and new results showing that all balanced k-colorings in the NNN case on square lattices are spatially periodic.

2000 *Mathematics Subject Classification.* Primary 37C10, 37C20; Secondary 34A34.
This research was supported in part by NSF Grant DMS-0244529 and ARP Grant 003652-0032-001. The work of FA was supported in part by a grant from FAPESP.

1 Coupled cell systems

A wide variety of physical and biological systems can naturally be modelled by networks of nonlinear dynamical systems, see Wang [7], Stewart [5]. The theoretical understanding of such systems is also under intensive development. Networks of differential equations possess additional structure, namely, canonical observables — the dynamical behavior of the individual nodes [3]. These observables can be compared, revealing such features as synchrony or specified phase-relations in periodic solutions, and these features are important in many applications. Any theoretical treatment of network dynamics must take this additional structure into account. In particular, the topology (or 'architecture') of the network imposes constraints on the dynamics, with the result that many new phenomena become 'generic' for a given architecture, see for example Golubitsky *et al.* [2].

Stewart, Golubitsky, Pivato, and Török [6, 4] formalize the concept of a *coupled cell network*, where a *cell* is a system of ordinary differential equations (ODEs) and a *coupled cell system* consists of cells whose equations are coupled. Stewart *et al.* define the architecture of coupled cell networks and develop a theory that shows how network architecture leads to synchrony. The architecture of a coupled cell network is a graph that indicates which cells have the same phase space, which cells are coupled to which, and which couplings are the same.

In its simplest form the *input set* $I(c)$ of a cell c consists of all cells coupled to c. Two input sets are *isomorphic* if there is a bijection between the input sets that preserves coupling types. A coupled cell network is *homogeneous* if the input sets of all cells are isomorphic. Suppose we color the cells in a homogeneous network. That coloring is *balanced* if for each pair of cells c and d with the same color, there is a color and coupling type preserving bijection from $I(c)$ to $I(d)$.

A lattice dynamical system is a homogeneous coupled cell system with cells indexed by a lattice $\mathcal{L}$. Such a system has the form

$$\dot{x}_c = g(x_c, x_{I(c)}) \qquad c \in \mathcal{L} \tag{1.1}$$

where $x_c \in \mathbf{R}^n$, $I(c) = \{c_1, \ldots, c_k\}$, $x_{I(c)} = (x_{c_1}, \ldots, x_{c_k}) \in (\mathbf{R}^n)^k$ and $g : (\mathbf{R}^n)^{k+1} \to \mathbf{R}^n$.

A *polydiagonal* is a subspace of the phase space of a coupled cell network that is defined by equality of cells coordinates. A polydiagonal is *robustly polysynchronous* if it is flow-invariant for every coupled cell system with the given network architecture. Robustly polysynchronous polydiagonals are identified with *patterns of synchrony*. Stewart *et al.* [6, Theorem 6.1] prove that a polydiagonal is robustly polysynchronous if and only if the coloring given by coloring cells that have the same coordinates with the same color is balanced. Thus, classifying robustly polysynchronous polydiagonals is equivalent to the combinatorial question of classifying balanced colorings.

Some patterns of synchrony can be predicted by symmetry, namely, those that correspond to fixed-point subspaces of the group of network symmetries. However, not all patterns can be obtained in this way, and some of these nonsymmetric patterns are quite interesting. Golubitsky *et al.* [2] give an infinite class of two-color patterns of synchrony on square lattice dynamical systems with nearest neighbor coupling. Wang and Golubitsky [8] classify all possible two-color patterns of synchrony of square lattice differential equations with two different architectures —

nearest neighbor coupling (NN) and both nearest neighbor and next nearest neighbor coupling (NNN). These classification results are stated in Theorems 3.1-3.2. It follows from these theorems that with NNN architecture balanced two-colorings are finite in number and spatially doubly-periodic. Thus, there is a profound difference between balanced two-colorings in the NN and NNN cases: one classification is finite, the other is infinite; one set has spatially periodic and nonperiodic colorings, the other has only periodic colorings.

In this note we prove that all balanced colorings of a one-dimensional lattice are spatially periodic (Theorem 2.4) and that for each $k \geq 3$ there are exactly four balanced k colorings (Theorem 2.5). The proof that balanced k colorings are periodic and finite in number uses a notion of *window* (a small finite patch of the network that determines the full pattern associated to any balanced k-coloring) and that proof generalizes to higher dimensions. In Section 3 we discuss balanced colorings in square arrays with nearest and next nearest neighbor coupling. We begin by summarizing the classification of balanced 2-colorings by Wang and Golubitsky [8]. Then we state the generalization of the results in Section 2 to square lattices and end with a brief discussion of some generalizations. Proofs and exact statements of these results appear in Antoneli *et al.* [1].

2 Linear arrays with nearest neighbor coupling

Let G be a one-dimensional lattice network with nearest neighbor coupling. An *interval* of a one-dimensional lattice network G is a finite sequence of consecutive cells $c_1 \ldots c_n$. The *interior* of this interval is the cell sequence $c_2 \ldots c_{n-1}$.

Definition 2.1 Given a one-dimensional lattice network G and a balanced k-coloring, we say that an interval is a *window* for the balanced coloring if its *interior* contains all k colors.

Note that the part of any balanced k-coloring that lies inside a window extends uniquely to the whole lattice. Since the k-coloring is balanced and each of the k colors has its nearest neighbors inside the interval, the colors of the cells on the right and left ends of the interval are determined. Proceeding recursively, the coloring can be extended uniquely to the whole lattice.

Lemma 2.2 *Given a balanced coloring of G with ℓ colors. Then every interval of length $2k-1$, where $k \leq \ell$, contains k colors.*

Proof Certainly every interval of length 1 contains one color. By induction, assume that the statement is valid for $k-1$. An interval of length $2k-1$ has the form $ac_1 \ldots c_{2k-3}b$. By induction the interval $c_1 \ldots c_{2k-3}$ contains at least $k-1$ colors. If that subinterval contains k colors, we are done. Similarly, if a or b is a kth color, then we are done. Suppose not. Since the coloring is balanced and each of the $k-1$ colors has its nearest neighbors inside the interval, the coloring uniquely extends to the whole lattice and the extended coloring has fewer than k colors, a contradiction. □

Proposition 2.3 *Any interval of size $2k+1$ is a window for every balanced k-coloring of the linear lattice.*

Proof This follows directly from the definition of window and Lemma 2.2. □

Theorem 2.4 *For every k there exists a finite number of balanced k-colorings in one-dimensional lattice dynamical systems with nearest neighbor coupling. Moreover, all balanced colorings are periodic.*

Proof First, given a window W, there is a finite number of ways to color the cells in W with k colors. In particular, there is a finite number of balanced k-colorings inside any window. Therefore, the number of balanced k-colorings on the lattice is finite.

Given a balanced k-coloring of G, by Proposition 2.3, any interval of size $2k+1$ is a window that determines it uniquely. Let W be one such window. Consider a covering of the lattice by disjoint translates of W. Since there is an infinite number of such translates but only a finite number of balanced colorings, there must be at least two translates W_1 and W_2 exhibiting exactly the same coloring and thus these two windows determine the same balanced k-coloring. Now observe that any translation of a balanced k-coloring is again a balanced k-coloring. It follows that the translation that takes W_1 to W_2 leaves the k-coloring invariant and so the coloring is periodic. □

Theorem 2.5 *Fix k and let $A_1, A_2, \ldots, A_k$ be k distinct colors. Then, every balanced k-coloring of the one-dimensional lattice with nearest neighbor coupling has one the following four forms.*

(i) No reflection; period k

$$\cdots \mathbf{A_1 A_2} \cdots \mathbf{A_k}\ \mathbf{A_1 A_2} \cdots \mathbf{A_k} \cdots$$

(ii) Two reflections both without fixed cells; period $2k$

$$\cdots \mathbf{A_1 A_2} \cdots \mathbf{A_k A_k} \cdots \mathbf{A_2 A_1}\ \mathbf{A_1 A_2} \cdots \mathbf{A_k A_k} \cdots \mathbf{A_2 A_1} \cdots$$

(iii) Two reflections one with fixed cell and one without fixed cell; period $2k-1$

$$\cdots \mathbf{A_1 A_2} \cdots \mathbf{A_k} \cdots \mathbf{A_2 A_1}\ \mathbf{A_1 A_2} \cdots \mathbf{A_k} \cdots \mathbf{A_2 A_1} \cdots$$

(iv) Two reflections both with fixed cells; period $2k-2$

$$\cdots \mathbf{A_1 A_2} \cdots \mathbf{A_k} \cdots \mathbf{A_2 A_1 A_2} \cdots \mathbf{A_k} \cdots \mathbf{A_2 A_1} \cdots$$

All of these balanced colorings are distinct when $k \geq 3$; balanced colorings (i) and (iv) are the same when $k = 2$; and all balanced colorings are identical when $k = 1$.

Proof The case $k = 1$ is straightforward and we assume $k \geq 2$. We fix a balanced k-coloring and let $m \geq 0$ be the smallest number of cells that occur between two cells of the same color. We claim that the balanced coloring has a reflection symmetry that fixes no cells when $m = 0$; a reflection that fixes a cell when $m = 1$; and is of type (i) when $m \geq 2$.

When $m = 0$ the balanced coloring contains two neighboring cells of the same color. Thus the balanced pattern contains $\cdots AA \cdots$. If color B is the left neighbor of the first A, then balanced implies that B is also the right neighbor of the second A and the pattern has the form $\cdots BAAB \cdots$. Inductively, balanced implies that the coloring has a reflection symmetry that fixes no cells.

When $m = 1$ the balanced coloring contains the pattern $\cdots ABA \cdots$. If color C is the left neighbor of the first A, then balanced implies that C is also the right neighbor of the second A and the pattern has the form $\cdots CABAC \cdots$. Inductively, balanced implies that the coloring has a reflection symmetry that fixes a cell.

When $m \geq 2$, the balanced coloring contains the pattern

$$\cdots AB_1B_2\cdots B_mA\cdots$$

where, from the definition of m, the colors of A and all of the B_j are distinct. It follows from balanced that the color to the left of the first A must be B_m and the color to the right of the second A must be B_1; that is, the pattern contains

$$\cdots B_mAB_1B_2\cdots B_mAB_1\cdots$$

Inductively, balanced implies that the pattern has the form (i) (and $m = k-1$).

We can now assume that the coloring has a reflection symmetry. All lattice reflections have the form

$$r_j(i) = j - i$$

where j is an integer. Note that r_j is the reflection about the point $\frac{j}{2}$ and is a reflection with a fixed cell when j is even and a reflection without a fixed cell when j is odd.

Theorem 2.4 states that every balanced k-coloring is periodic and hence has a translation symmetry $i \mapsto i + p$ where p is the minimum period of the coloring. It follows that the balanced pattern must have two independent reflections; just compose translation by p with r_j to obtain

$$r_j(i+p) = j - (i+p) = r_{j-p}(i)$$

The point of reflection in the second reflection is distance $\frac{p}{2}$ from the point of reflection of the first. Moreover, the composition of two reflections r_j and r_q is a translation by $j - q$; just compute

$$r_j(r_q(i)) = r_j(q-i) = j - (q-i) = i + (j-q)$$

It follows that a pattern with minimum period p cannot have reflections whose points of reflection are closer than $\frac{p}{2}$. Hence these patterns have exactly two reflections in a minimum period. Note also that if p is even either both reflections have a fixed cell or both do not. If p is odd there is one reflection of each type. Thus, there are three cases to consider: p even and reflections with a fixed cell; p even and reflections without fixed cells; and p odd and one reflection of each type. We claim that these cases lead to the patterns (iv), (ii), and (iii), respectively.

We now consider the case where p is even and the coloring has two reflections without fixed cells and show that this case leads to case (ii). It follows by reflection about the midpoint between two A cells, as in the case $m = 0$, that the pattern has the form

$$\cdots B_0B_0B_1\cdots B_\ell B_{\ell+1}B_{\ell+1}\cdots$$

where $\ell = \frac{p}{2} - 2$. Reflecting about $B_{\ell+1}$ leads to the periodic pattern

$$\cdots B_0B_0B_1\cdots B_\ell B_{\ell+1}B_{\ell+1}B_\ell\cdots B_1B_0B_0\cdots$$

We claim that balanced implies that the colors are $B_0, \ldots, B_{\ell+1}$ are all distinct. Suppose that B_{j+1} is the first color in this sequence that is the same as one of the colors $B_0, \ldots, B_j$. If the color of B_{j+1} is the same as the color of B_j, then, as in the case $m = 0$ above, the pattern has an additional reflection. If the color of B_{j+1} is the same as the color of B_{j-1}, then, as in the case $m = 1$ above, the pattern has an additional reflection. If the color of B_{j+1} is the same as one of the previous colors, then the coloring is not balanced. The claim is verified and the number of colors in this pattern is $k = \ell + 2$ and the period is $p = 2\ell + 4 = 2k$, as desired.

The remaining two cases are proved similarly by considering the appropriate reflections. □

3 Balanced colorings on square arrays

In this section we review results on balanced colorings on square arrays. In particular we note that balanced two-colorings on square arrays need not be spatially periodic when only nearest neighbor coupling is assumed, whereas they are spatially periodic if both nearest and next nearest neighbor coupling are assumed. Indeed, the results of Section 2 can be extended to square arrays; namely, balanced k-colorings are spatially periodic when nearest and next nearest neighbor coupling are assumed.

Theorem 3.1 ([8]) *There are eight two-color periodic patterns of synchrony of square lattice differential equations with nearest neighbor coupling shown in Figure 1. There are two infinite families of two-color patterns of synchrony that are generated from the periodic patterns in Figure 2 by interchanging black and white on diagonals along which black and white cells alternate. Up to symmetry, these are all of the two-color patterns of synchrony.*

Figure 1 Illustrations of patterns of synchrony of finite classes [8].

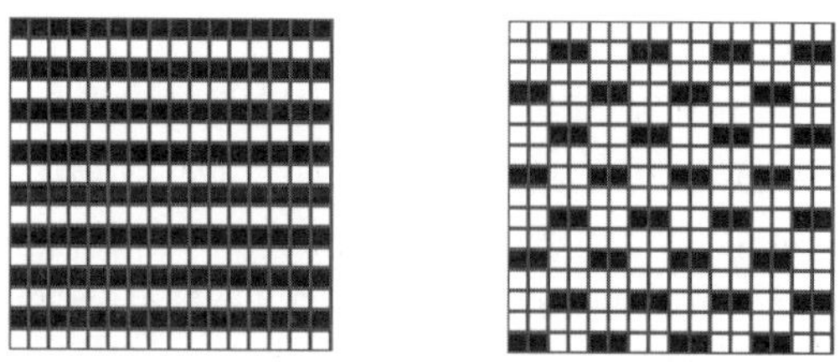

Figure 2 Illustrations of patterns of synchrony of infinite classes [8].

Some examples of the infinite number of nonperiodic balanced two-colorings are given in Figure 3.

Figure 3 Patterns of synchrony that are not spatially periodic [2, 8].

Theorem 3.2 ([8]) *Up to symmetry, there are twelve two-color patterns of synchrony in square lattice differential equations with nearest and next nearest neighbor couplings: the seven patterns in Figure 1 except for the boxed pattern, Figure 2 (left), and the four patterns in Figure 4.*

Figure 4 The four patterns in this figure and Figure 2 (left) are NNN-balanced [8].

Remark 3.3 All twelve patterns are doubly-periodic. Those patterns in Figure 4 can be generated from Figure 2 (a) by interchanging black and white along diagonals on which black and white cells alternate.

Indeed more is true once next nearest neighbor coupling is assumed.

Theorem 3.4 ([1]) *For any integer $k > 0$, there are only a finite number of k-color patterns of synchrony of square lattice dynamical systems with nearest and next nearest couplings. All of these patterns are periodic.*

The proof of Theorem 3.4 proceeds as in the case of one-dimensional arrays. Given a k-coloring of a two-dimensional square lattice network G, a bounded square subarray of G is called a *window* if its interior contains all k colors. We then prove that windows determine uniquely the whole pattern and that every square of size $2k+1$ is a window. The proof of the theorem follows from these observations.

We call a lattice *Euclidean* if it is generated by vectors of the same length. It is true that if enough different kinds of coupling are present, then all balanced k-colorings of any planar Euclidean lattice dynamical system are spatially periodic. See [1] for details. We conjecture that this result is valid in all dimensions.

References

[1] F. Antoneli, A.P.S. Dias, M. Golubitsky and Y. Wang. Patterns of synchrony in lattice dynamical systems. *Nonlinearity* **18** (2005) 2193–2209.

[2] M. Golubitsky, M. Nicol, and I. Stewart. Some curious phenomena in coupled cell networks. *J. Nonlinear Sci.* **14**(2) (2004) 119–236.

[3] M. Golubitsky and I. Stewart. *The Symmetry Perspective: From Equilibrium to Chaos in Phase Space and Physical Space*, Progress in Mathematics **200**, Birkhäuser, Basel 2002.

[4] M. Golubitsky, I. Stewart and A. Török. Patterns of synchrony in coupled cell networks with multiple arrows. *SIAM J. Appl. Dynam. Sys.* **4**(1) (2005) 78-100.

[5] I. Stewart. Networking opportunity, *Nature* **427** (2004) 601–604.

[6] I. Stewart, M. Golubitsky, and M. Pivato. Symmetry groupoids and patterns of synchrony in coupled cell networks. *SIAM J. Appl. Dynam. Sys.* **2 (4)** (2003) 609–646.

[7] X.F. Wang. Complex networks: topology, dynamics and synchronization, *Internat. J. Bif. Chaos* **12** (2002) 885–916.

[8] Y. Wang and M. Golubitsky. Two-color patterns of synchrony in lattice dynamical systems. *Nonlinearity* **18** (2005) 631–657.

Fields Institute Communications
Volume **49**, 2006

Waves in Extended Systems: From Low- to High-dimensional Behavior

O. Batiste
Departament de Física Aplicada
Universitat Politècnica de Catalunya
08034 Barcelona, Spain
`oriol@fa.upc.edu`

E. Knobloch
Department of Physics
University of California
Berkeley, CA 94720, USA
`knobloch@physics.berkeley.edu`

This paper is dedicated to William F. Langford

Abstract. Experiments and simulations near onset of convection in ^{3}He-^{4}He mixtures with a negative separation ratio in containers of moderately large aspect ratio Γ reveal a large variety of wave-like states with complex time-dependence. Remarkably, many of these states can be understood using a simple set of odes derived from normal form theory. The reason for this success is a consequence of the small number of active degrees of freedom present when $(R - R_c)/R_c \ll \Gamma^{-2}$, where R is the Rayleigh number and R_c is its critical value for the primary Hopf bifurcation. The transition regime $(R - R_c)/R_c \sim \Gamma^{-2}$ involving more degrees of freedom is also explored. Among the new states in this regime are convectons: states of spatially localized steady convection.

1 Introduction

Convection in binary fluid mixtures with a negative separation ratio typically sets in via a subcritical Hopf bifurcation. In extended domains this bifurcation leads to a variety of traveling wave states very close to the onset of the instability [46, 34, 45, 31]. In rectangular containers these include the so-called *blinking states* [34, 45] and the *repeated transients* [31]. In such containers translation invariance is absent, and the eigenmodes of the linear stability problem are either odd or even under left-right reflection [14]. Recent work [35, 6, 5, 4] shows that a description of

---

2000 *Mathematics Subject Classification.* Primary 54C40, 14E20; Secondary 46E25, 20C20.
The second author was supported in part by NSF Grant DMS-0305968.

the observed dynamics based on amplitude equations for such modes succeeds in capturing the behavior observed in both experiments and numerical simulations remarkably well. We describe these developments below.

When a mixture of two miscible fluids is heated the heavier component in general migrates towards the colder boundary. However, in mixtures with a negative separation ratio the opposite occurs. Such a mixture responds to heating from below by setting up a competing stabilizing concentration gradient. When this is strong enough the initial instability is a Hopf bifurcation [28]. The original experiments on this system were performed in salt-water mixtures [9], although subsequent experiments focused on ^{3}He-^{4}He mixtures [46, 1, 47, 8] and water-ethanol mixtures [34, 45, 31]. Of these the latter readily permit flow visualization, a fact invaluable for the interpretation of all measurements. However, the recent development of visualization techniques at cryogenic temperatures [49] offers the possibility that comparable experiments can now be performed on ^{3}He-^{4}He mixtures.

Existing simulations of two-dimensional binary fluid convection in rectangular containers [5] are restricted to water-ethanol mixtures in containers of dimensionless aspect ratio $16 \leq \Gamma \leq 17$ and $\epsilon \equiv (R-R_c)/R_c \sim 10^{-3}$ or less, as in the experiments of Kolodner [31]. Here R_c is a dimensionless measure of the critical temperature (the critical Rayleigh number) for the primary Hopf bifurcation. The simulations employ the experimental parameter values and boundary conditions, and reproduce much of the behavior reported in the experiments. In particular they demonstrate that:

- The experiments can be understood on the basis of *two*-dimensional simulations.
- In the regime $|\epsilon| \ll \Gamma^{-2}$ the behavior exhibited by the partial differential equations (and hence the experiments) is fundamentally low-dimensional even though the aspect ratio Γ may be quite large.
- The primary instability takes the form of a subcritical bifurcation to an unstable time-periodic *chevron* state, consisting of traveling waves that propagate in opposite directions, usually outward from the center, in the two halves of the container. Such states are either odd or even under left-right reflection, and may acquire stability at finite amplitude via a saddle-node bifurcation.
- Stable blinking states are two-frequency states and set in when stable chevrons lose stability at a symmetry-breaking Hopf bifurcation. The blinking amplitude increases with ϵ, and the blinking may become nonperiodic.
- In cases where the secondary Hopf bifurcation occurs on the unstable chevron branch below the saddle-node bifurcation the first observed state is the repeated transient state. This state is a three-frequency state, and may set in *without observable* hysteresis at $R = R_c$.
- With increasing ϵ the lowest frequency in the repeated transient state increases but its contribution decreases, and may disappear at a (typically hysteretic) Hopf bifurcation, leaving a large amplitude blinking state.
- Regular blinking states are observed near onset only for aspect ratios differing roughly by $\pi/k_c \approx 1$, corresponding to mode interaction points. Here k_c is the linear theory wave number.

In this Chapter we explain these conclusions and illustrate them using simulations of ^{3}He-^{4}He mixtures in somewhat smaller domains. This choice reduces the extremely long transients that plagued the water-ethanol simulations, and permits us to investigate the above questions in more detail than would otherwise be possible. Some of our results are reported in [3, 4]. The results show that the theory developed in [5] for $|\epsilon| \ll \Gamma^{-2}$ extends to low Prandtl number mixtures, identify *chaotic* repeated transients, and illustrate the dramatic changes that occur when $|\epsilon| \sim \Gamma^{-2}$.

2 The governing equations

We consider a binary mixture in a two-dimensional rectangular container $D \equiv \{x, z | 0 \le x \le \Gamma, -\frac{1}{2} \le z \le \frac{1}{2}\}$ heated uniformly from below, and nondimensionalize the equations using the depth of the layer as the unit of length and t_d, the thermal diffusion time in the vertical, as the unit of time. In the Boussinesq approximation appropriate to the experiments the resulting equations take the form [5]

$$\mathbf{u}_t + (\mathbf{u} \cdot \nabla)\mathbf{u} = -\nabla P + \sigma R[\theta(1+S) - S\eta]\hat{\mathbf{z}} + \sigma \nabla^2 \mathbf{u}, \tag{2.1}$$

$$\theta_t + (\mathbf{u} \cdot \nabla)\theta = w + \nabla^2 \theta, \tag{2.2}$$

$$\eta_t + (\mathbf{u} \cdot \nabla)\eta = \tau \nabla^2 \eta + \nabla^2 \theta, \tag{2.3}$$

together with the incompressibility condition

$$\nabla \cdot \mathbf{u} = 0. \tag{2.4}$$

Here $\mathbf{u} \equiv (u, w)$ is the velocity field in (x, z) coordinates, P is the pressure, and θ denotes the departure of the temperature from its conduction profile, in units of the imposed temperature difference ΔT. The variable η is defined such that its gradient represents the dimensionless mass flux. Thus $\eta \equiv \theta - \Sigma$, where $C \equiv -z + \Sigma$ is the concentration of the heavier component in units of the concentration difference that develops across the layer as a result of cross-diffusion. The system is specified by four dimensionless parameters: the Rayleigh number R providing a dimensionless measure of the imposed temperature difference ΔT, the separation ratio S that measures the concentration contribution to the buoyancy force due to cross-diffusion, and the Prandtl and Lewis numbers σ, τ, in addition to the aspect ratio Γ.

To model the experiments we take the boundaries to be no-slip everywhere, with the temperature fixed at the top and bottom and no sideways heat flux. The final set of boundary conditions is provided by the requirement that there is no mass flux through any of the boundaries. The boundary conditions are thus

$$\mathbf{u} = \mathbf{n} \cdot \nabla \eta = 0 \text{ on } \partial D, \tag{2.5}$$

and

$$\theta = 0 \text{ at } z = \pm 1/2, \qquad \theta_x = 0 \text{ at } x = 0, \Gamma. \tag{2.6}$$

Here ∂D denotes the boundary of D.

Equations (2.1)-(2.6) are equivariant with respect to the operations

$$R_x : \qquad (x, z) \to (\Gamma - x, z), \qquad (\psi, \theta, \eta) \to (-\psi, \theta, \eta), \tag{2.7}$$

$$\kappa : \qquad (x, z) \to (x, -z), \qquad (\psi, \theta, \eta) \to (-\psi, -\theta, -\eta), \tag{2.8}$$

where $\psi(x, z, t)$ is the streamfunction, defined by $(u, w) = (-\psi_z, \psi_x)$. These two operations generate the symmetry group D_2 of a rectangle. It follows that even solutions, i.e., solutions invariant under R_x, satisfy $(\psi(x,z),\ \theta(x,z),\ \eta(x,z)) = (-\psi(\Gamma - x, z),\ \theta(\Gamma - x, z),\ \eta(\Gamma - x, z))$ at each instant in time, while odd solutions are invariant under κR_x and satisfy $(\psi(x,z),\ \theta(x,z),\ \eta(x,z)) = (\psi(\Gamma - x, -z),$

$-\theta(\Gamma - x, -z)$, $-\eta(\Gamma - x, -z))$, again at each instant of time. At midlevel, $z = 0$, the latter therefore satisfy $(\psi(x, 0), \theta(x, 0), \eta(x, 0)) = (\psi(\Gamma - x, 0), -\theta(\Gamma - x, 0), -\eta(\Gamma - x, 0))$, and so are odd in the conventional sense.

3 Linear theory

To determine the critical value R_c of the Rayleigh number at which the conduction state $\psi = \theta = C = 0$ loses stability to overstable convection and the corresponding frequency ω_c we look for solutions to the linearized equations of the form $f(x, z)e^{(s+i\omega)t}$. The condition $s = 0$ defines the onset of instability and yields a complex condition that can be solved for $R \equiv R_c$ and $\omega \equiv \omega_c$ as a function of the aspect ratio Γ for various values of σ, τ and the separation ratio S [6].

The solution of this problem for the parameters $\sigma = 0.6$, $\tau = 0.03$ appropriate for the experiments [46] indicates that the competition between odd and even modes in this system takes one of two basic forms, depending on the separation and aspect ratios (see fig. 1). When $|S|$ is small (i.e., close to $|S_{TB}|$, the Takens-Bogdanov point) and Γ not too large the mode interaction takes the form familiar from studies of stationary Rayleigh-Bénard convection with non-Neumann boundary conditions [16, 40, 21]: the neutral curves $R_c(\Gamma)$ divide neatly between different families with no intermingling among them. Each family consists of a pair of braided neutral curves, one for an odd mode and the other for an even mode, with each family well separated from the next, at least for the low-lying families. The crossings between odd and even modes within each family are structurally stable because of their different parity. For the case of interest here, i.e., Γ and $|S|$ large enough, the situation is quite different. There are now no distinct families of neutral curves and all modes (including like-parity modes) cross. In general these mode crossings are all structurally stable, either because the modes have opposite parity, or because their frequencies at the mode crossing are nonresonant. Figure 1 shows the location of the transition between these types of behavior in the (Γ, S) plane.

Figure 2 shows the eigenvalues $R_c(\Gamma)$ and $\omega_c(\Gamma)$ for two values of the separation ratio, $S = -0.001$ and $S = -0.01$, in the range $4 \leq \Gamma \leq 12$. For $S = -0.001$ fig. 2 reveals an oscillatory approach of both sets of curves towards the critical Rayleigh number R_∞ and ω_∞ for an unbounded domain. As already mentioned the braiding of the neutral stability curves $R(\Gamma)$ seen in fig. 2a is a consequence of the small value of $|S|$; because of it the neutral stability curves for the first odd and even modes cross repeatedly. Such mode crossings indicate the presence of codimension-two bifurcations. Figure 2b shows that at these points the frequencies of the competing modes are distinct. Consequently these mode interaction points correspond to *nonresonant* double Hopf bifurcations. However, when $S = -0.01$ the situation changes: for large enough aspect ratios ($\Gamma > 10$) the two lowest neutral curves develop cusps (fig. 2c). The presence of these cusps is reflected in the discontinuous jumps in the corresponding frequency curves (fig. 2d). Detailed calculations [6] reveal that these cusps form once neutral stability curves corresponding to like parity modes from adjacent families start to cross. These crossings are mediated by double Hopf bifurcations with 1:1 resonance [48] located at discrete points $(R_c^{(3)}, \Gamma_c^{(3)}, S_c^{(3)})$ in the three-dimensional parameter space (R, Γ, S), cf. fig. 1.

In fig. 3 we show the bifurcating eigenvectors for $\Gamma = 10$ when $S = -0.001$ (left panels) and $S = -0.01$ (right panels). Both modes are even and are represented in the form of space-time diagrams, with time increasing upwards. When $S = -0.001$

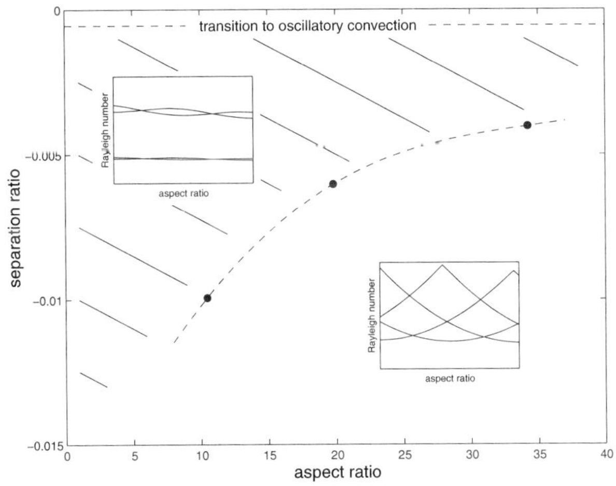

Figure 1 The (Γ, S) plane for ^{3}He-^{4}He parameters showing the approximate location of mode avoidance (hatched region) and of mode crossing (unhatched region).

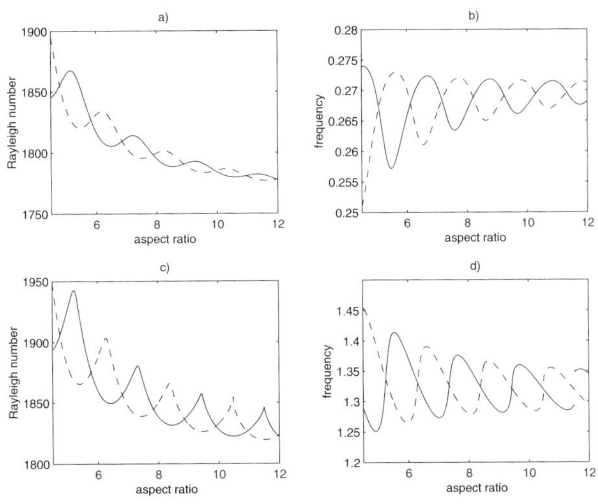

Figure 2 Onset of convection in ^{3}He-^{4}He mixtures ($\sigma = 0.6$, $\tau = 0.03$) in moderate aspect ratio containers. (a) The critical Rayleigh number R_c and (b) the corresponding frequency ω_c for the first even (solid line) and odd (dashed line) mode as a function of the aspect ratio Γ for $S = -0.001$. (c,d) The same but for $S = -0.01$.

the eigenvector takes the form of a standing wave, with the dynamics at the two sidewalls in phase. The amplitude of the standing oscillations peaks in the middle of the container and decreases towards the sidewalls. There is a considerable phase lag between the temperature and concentration oscillation, a consequence of the small

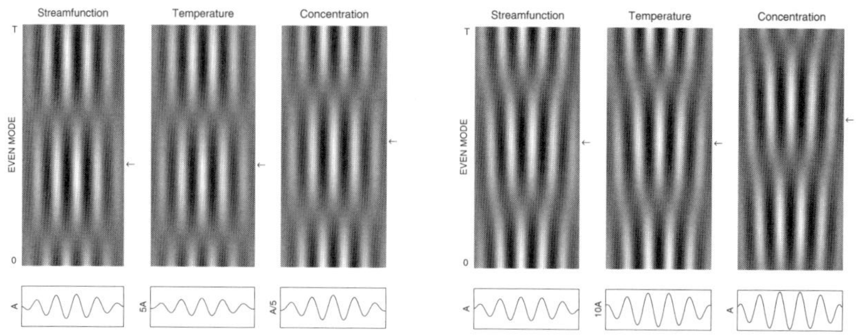

Figure 3 The eigenvector (ψ, θ, Σ) of the linear stability problem at $z = 0$ for $\Gamma = 10$, $\sigma = 0.6$, $\tau = 0.03$, $S = -0.001$, $R_c = 1784.09$, $\omega_c = 0.267$ (left panels), and $S = -0.01$, $R_c = 1827.72$, $\omega_c = 1.35$ (right panels), shown in the form of space-time diagrams with $0 \leq x \leq \Gamma$ horizontally and time increasing upwards. The solutions are sinusoidal with period $T \equiv 2\pi/\omega_c$. In the former case the eigenfunctions are standing waves but become 'chevron'-like in the latter. The amplitude variation across the domain is indicated below each space-time plot.

value of τ. As S decreases the eigenvector gradually develops into a 'chevron'-like state, consisting of waves propagating outward from the container center in such a way that the reflection symmetry in $x = \Gamma/2$ is preserved at all times, as in fig. 3. In contrast an odd parity chevron (at $z = 0$) is at all times odd with respect to this reflection. Note that despite appearances these solutions are *strictly sinusoidal* in time: the periodic defect formation at $x = \Gamma/2$ arises because the eigenfunction ψ is a superposition of four functions each of which has the form $\psi_j \exp i(\omega_c t \pm k_j x)$, $j = 1, \ldots, 4$, with the k_j possibly complex. Each of these functions describes waves propagating with (local) phase velocity $\pm\omega_c/\text{Re}k_j$. In the bulk the eigenfunction is dominated by the largest contribution; this contribution has a real wavenumber and describes oscillations that are almost standing. However, when the time-dependent amplitude of this component passes through zero (which occurs twice per period) the remaining contributions briefly reveal themselves. In the eigenfunction shown the largest of these has a relatively large phase velocity, and is responsible for the episodic propagation that is so characteristic of the eigenfunctions shown. In both panels of fig. 3 the dominant local wavenumber remains remarkably uniform across the cell despite the nonuniformity of the amplitude of the eigenfunction. The sources (or sinks) described by these *linear* eigenfunctions persist into the nonlinear regime (Section 4), indicating that in the present system these defects have a nontopological origin.

Equations (2.1)-(2.6) are solved as in [5]. Typically we used 170 collocation points in the x-direction and 30 collocation points in the z-direction, with a time step of $10^{-3}t_d$. We use the vertical velocity at the points $(x, z) = (0.13\Gamma, 0)$ (near the left sidewall) and $(0.87\Gamma, 0)$ (near the right sidewall) as a proxy for shadowgraph intensity measurements. Moreover, monitoring point quantities at mirror locations suffices to determine the spatial symmetry properties of the various possible time-dependent states. Below we describe our results for (a) $S = -0.1$ and (b) $S = -0.5$, with $\sigma = 0.06$, $\tau = 0.03$ and $\Gamma = 10$.

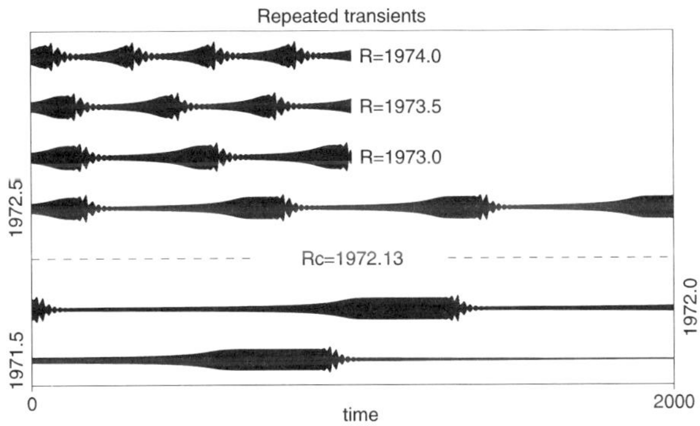

Figure 4 The repeated transient state in a ^{3}He-^{4}He mixture with $S = -0.1$, $\sigma = 0.6$, $\tau = 0.03$ and $\Gamma = 10$ at several values of R. The time series $w(x = 0.87\Gamma, z = 0, t)$ suggest the presence of three frequencies, with the lowest frequency ω_3 increasing from zero as R increases from $R \approx 1971.5$ ($\epsilon = \epsilon^* \approx -3.2 \times 10^{-4}$).

4 Extended states

4.1 When $S = -0.1$ linear theory shows that $R_c = 1972.13$, $\omega_c = 4.918$. The primary bifurcation is to an even chevron, and is substantially subcritical, cf. [10]. Figure 4 shows that the first stable nonlinear state takes the form of a *repeated transient*. States of this type were studied in detail by Kolodner [31] in experiments on water-ethanol mixtures, and their origin is discussed in [5]. The figure suggests that these states are *three-frequency* states, in which ω_1 is the fast chevron frequency, ω_2 represents the blinking frequency, while the third frequency ω_3 represents the slow modulation frequency. Figure 4 shows that these states consist of long episodes of slowly growing (even) chevrons. However, instead of saturating, this state becomes unstable to the onset of blinking, which leads to a collapse of the state to a small amplitude chevron, followed by a slow regrowth. In the time series shown these collapse events are periodic with period $T_3 = 2\pi/\omega_3$. Figure 5a shows that as R decreases the period $T_3 = 2\pi/\omega_3$ increases rapidly and apparently diverges at $R \approx 1971.5$; fig. 5b shows the corresponding Nusselt number, a dimensionless measure of the total heat flux across the layer. Such a divergence suggests that the three-frequency states are created via a global bifurcation. Since no stable nonzero solutions are present for smaller values of R for these parameter values the three-frequency repeated transient state represents the *first* nontrivial state of the system, much as occurs in water-ethanol mixtures [31, 5]. This interesting property of the system is explained in Section 5.

In contrast, as R increases the modulation frequency ω_3 increases but drops out from the time series between $R = 1980.0$ and $R = 1981.0$ (fig. 6). We identify this transition with a Hopf bifurcation (more precisely, a torus bifurcation), and note that fig. 6 indicates that this bifurcation is *supercritical*, i.e., viewed in the direction of decreasing R this bifurcation creates a *stable* three-frequency state from a *stable* two-frequency state, with no observable hysteresis. Figure 7 shows that this two-frequency state is a symmetric blinking state, and traces the evolution of this state towards larger values of R. The figure shows that at $R = 1995$ the blinking has

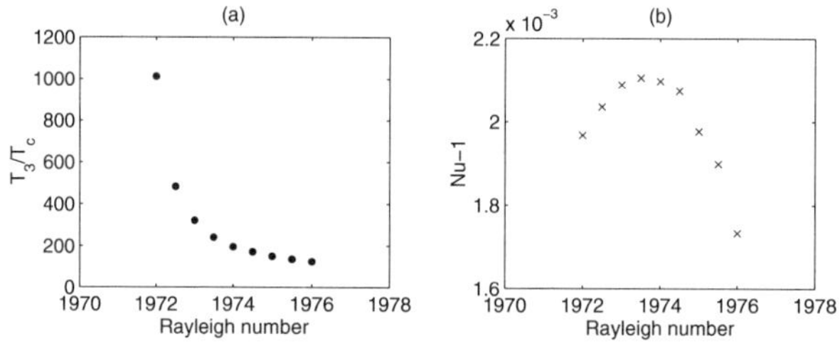

Figure 5 (a) The modulation period $T_3 \equiv 2\pi/\omega_3$ of the repeated transients when $S = -0.1$, $\sigma = 0.6$, $\tau = 0.03$, $\Gamma = 10$ in units of $T_c \equiv 2\pi/\omega_c$ as a function of the Rayleigh number R. (b) The corresponding Nusselt number.

become asymmetric and nonperiodic, while the time series for $R = 2000$ and $R = 2005$ may be periodic, but are strongly asymmetric. In contrast, when $R = 2010$ the blinking becomes once again periodic and symmetric. None of these transitions appear to be hysteretic. At yet larger values of R these states evolve into a state called a *fish* state [34], followed by a transition to stationary convection. Figure 8 shows an example of the fish state just prior to this transition. An examination of the spatial structure of the waves (fig. 9) shows that there is a brief phase of the oscillation during which small amplitude counterprogagating waves fill the container. This state is highly unstable, however, with the waves at one end growing at the expense of those at the other. Once the amplitude of the growing state is large enough the system shifts into a new and larger amplitude state in which the waves are spatially confined towards one side, with *no* waves at the other (see figs. 9 and 10a). This transition is marked by a dramatic drop in the frequency ω_1 since the waves in this pulse-like state travel much more slowly. We surmise that this is due to the large amplitude of this state. This change in frequency in turn increases the Nusselt number, and does so despite the fact that the waves no longer fill the whole domain. As time proceeds this confined state drifts in the direction of the waves within it, and settles next to the boundary, forming a wall-attached state (see figs. 9 and 10b). During the attachment process the localized state contracts and its amplitude decreases, presumably due to increased dissipation. As a result ω_1 increases again. During this process $w(x = 0.87\Gamma, z = 0, t)$ actually increases as activity moves to the region where w is measured. However, integral quantities such as the Nusselt number actually decrease, since the amplitude of the state falls on contact with the wall. The appearance of the prominent shoulder in the Nusselt number time series (fig. 8a) coincides with the formation of the wall-attached state. The wall-attached state is not stable, however, and continues to decrease in amplitude and contract until it becomes so weak and confined that waves start to regrow at the other sidewall. At this point the wall-attached state disintegrates, and the small amplitude extended chevron-like state is restored. All these transitions are quite easily distinguished in the time trace shown in fig. 8 and in the space-time plot of $\theta(x, 0, t)$ shown in fig. 9, and suggest that the formation of the initial pulse-like state is triggered by a nonlinear focusing effect reminiscent of the nonlinear Schrödinger equation.

Many of the above details resemble behavior observed in experiments on water-ethanol mixtures in [24], fig. 1, although in some cases [29] the frequency ω_1 *drops*

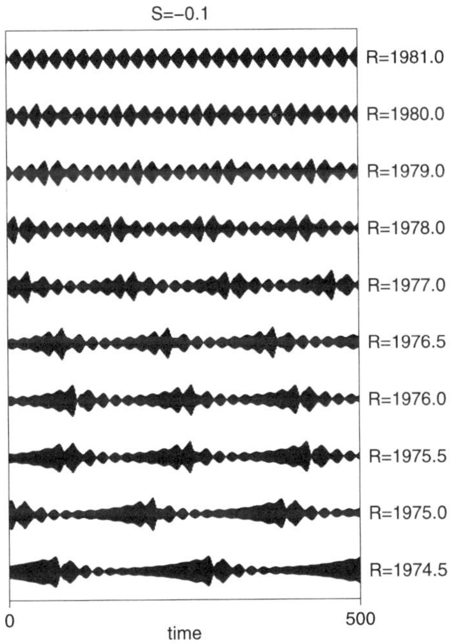

Figure 6 The time series $w(x = 0.87\Gamma, z = 0, t)$ for a ^{3}He-^{4}He mixture with $S = -0.1$, $\sigma = 0.6$, $\tau = 0.03$ and $\Gamma = 10$ at several values of R showing the transition from the repeated transient state in fig. 4 to a symmetric periodic blinking state at $R = 1981$ ($\epsilon = 4.5 \times 10^{-3}$).

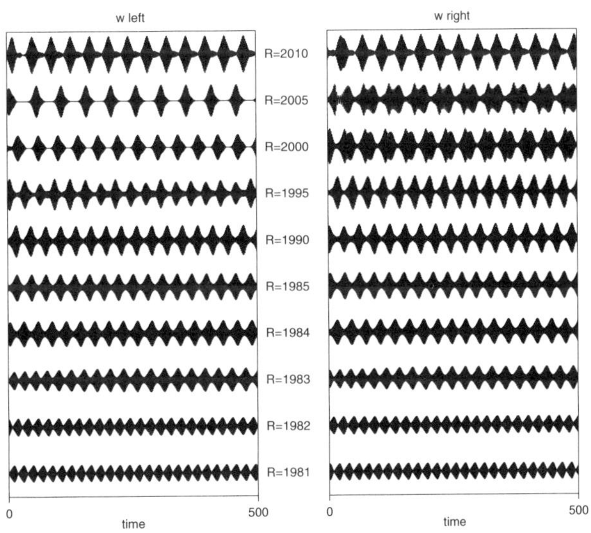

Figure 7 Time series $w(x = 0.13\Gamma, z = 0, t)$ (left column) and $w(x = 0.87\Gamma, z = 0, t)$ (right column) for different values of R increasing upwards, and $S = -0.1$, $\sigma = 0.6$, $\tau = 0.03$, $\Gamma = 10$.

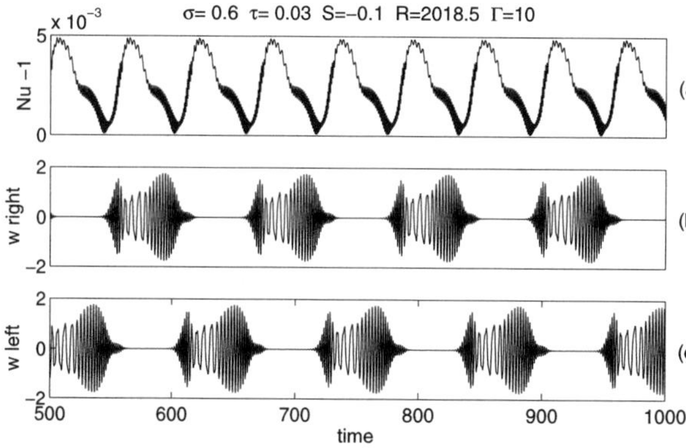

Figure 8 (a) The Nusselt number Nu(t), (b) $w(x = 0.87\Gamma, z = 0, t)$, (c) $w(x = 0.13\Gamma, z = 0, t)$ for $R = 2018.5$ ($\epsilon = 0.024$), and $S = -0.1$, $\sigma = 0.6$, $\tau = 0.03$, $\Gamma = 10$.

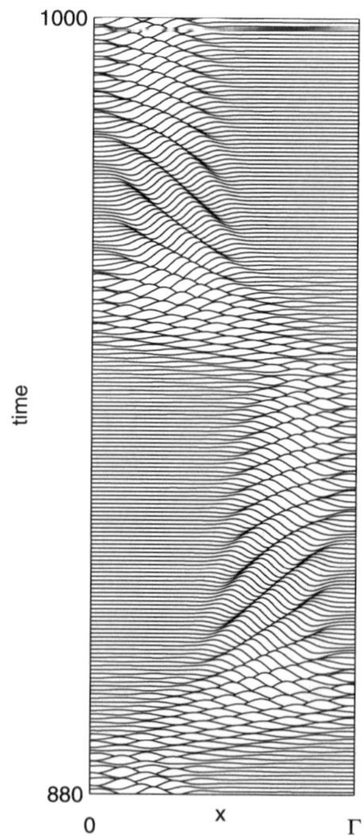

Figure 9 Space-time plot of the fish state in fig. 8 with time increasing upwards. The state is periodic in time with period T, and is invariant under the spatial reflection R_x followed by evolution for a time $T/2$.

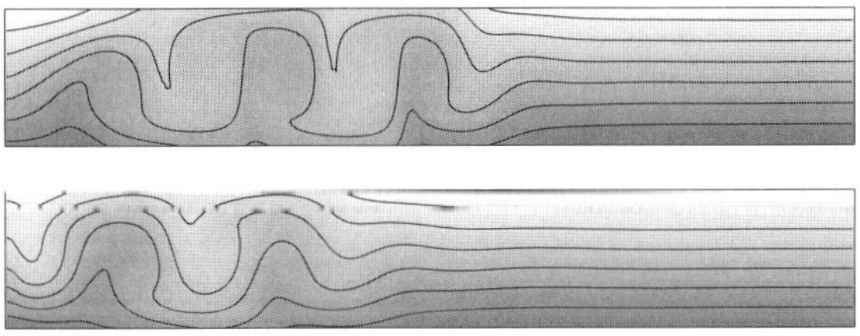

Figure 10 Contours of $C(x, z)$ during (a) the low frequency part of the fish state in fig. 8, and (b) during the high frequency phase that follows it. The localized state in (a) settles against the left sidewall forming temporarily the wall-attached left-traveling wave shown in (b).

as the confined state settles near the lateral wall. It is possible that this is because in this experiment the pulse drifts in a direction *opposite* to the waves within it (anomalous nonlinear dispersion). Experiments suggest that this occurs when $\epsilon < 0$ [41], while in our simulations $\epsilon > 0$.

4.2 When $S = -0.5$ ($R_c = 2643.43$, $\omega_c = 12.836$) the primary instability is again to an even chevron, but this time we find strongly irregular dynamics already quite close to onset (fig. 11). The time series in fig. 11 is best described as an intermittent repeated transient, in which the final collapse event may be preceded by several spatially symmetric bounces before the onset of the symmetry-breaking instability that disrupts the state and leads to the temporary formation of a confined state towards one side, much as already described for $S = -0.1$. This state then drifts towards the nearest wall and shrinks in lateral extent, until it triggers another collapse event that permits waves to grow at the other sidewall. The decaying symmetric blinking state that results reestablishes a small amplitude chevron state which then regrows on a much longer timescale. The spatially symmetric bounces are associated with relatively sharp peaks in the Nusselt number, while the symmetry-breaking collapse events produce bursts in the Nusselt number that are markedly asymmetric, much as in the fish state discussed above.

The frequency of the burst-like events in the Nusselt number increases with R (fig. 12). A periodic sequence of bursts, produced by a symmetric albeit complex state, is shown in fig. 13. Perhaps the most remarkable time series of all is shown in fig. 14 for $R = 2750$ (see also the corresponding space-time diagram in fig. 15). The time series apparently shows an irregular switching between two states, a large amplitude state with a relatively low ω_1, and a small amplitude state with a large ω_1. The former is a spatially confined slowly drifting wave, while the latter is an extended more-or-less symmetric pulsating chevron-like state. The latter is unstable to a symmetry-breaking blinking instability which amplifies the waves near one of the sidewalls at the expense of those near the opposite wall. The amplified waves are then reflected from the wall but continue to grow, slowing down markedly. This asymmetric state then abruptly collapses (by a mechanism that remains unclear) into a highly nonlinear spatially confined state consisting of slowly traveling waves. These waves propagate in the direction of the original reflected wave, and speed up as they approach the opposite wall. The whole pulse gradually retracts towards that wall, ultimately leaving much of the cell free of convection. Once the peak of

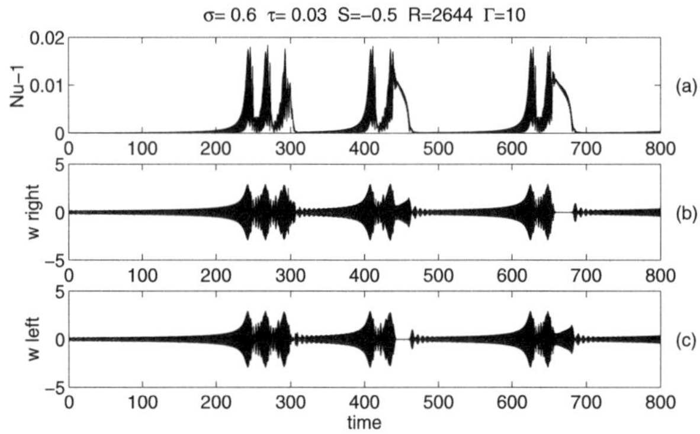

Figure 11 (a) The Nusselt number Nu(t), (b) $w(x = 0.87\Gamma, z = 0, t)$, (c) $w(x = 0.13\Gamma, z = 0, t)$ for $R = 2644$ ($\epsilon = 2.1 \times 10^{-4}$), and $S = -0.5$, $\sigma = 0.6$, $\tau = 0.03$, $\Gamma = 10$, showing irregular bursts.

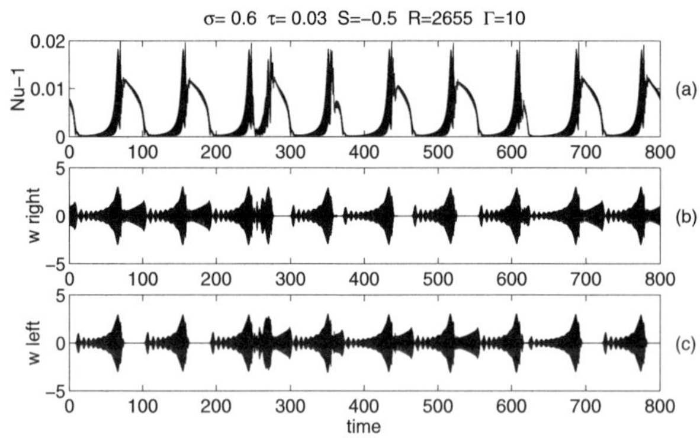

Figure 12 As for fig. 11 but showing the time series corresponding to $R = 2655$ ($\epsilon = 4.4 \times 10^{-3}$).

the pulse reaches the wall its amplitude drops dramatically and it disintegrates into smaller amplitude counterpropagating waves that invade the convection-free part of the container, permitting the regrowth of the original higher frequency small amplitude chevron-like state filling the domain (fig. 15). Overall this behavior resembles that observed for $S = -0.1$ (see figs. 9 and 10) but here it is much more dramatic, and we may think of it as back-and-forth 'sloshing'. The resulting state is reminiscent of a state found earlier in doubly-diffusive convection [17]; see also [37].

Figures 16 and 17 show that at these values of the Rayleigh number there is a wide variety of coexisting time-dependent states. The former shows a stable *odd* parity chevron state with a superposed small amplitude temporal modulation that coexists with the state shown in fig. 11 that is based on an even chevron. Figure 16 shows that the modulation is exactly out of phase in the two halves of the domain, and although it has a complex waveform it appears to be periodic in time. This state is therefore a symmetric blinking state, and indeed with decreasing R one finds

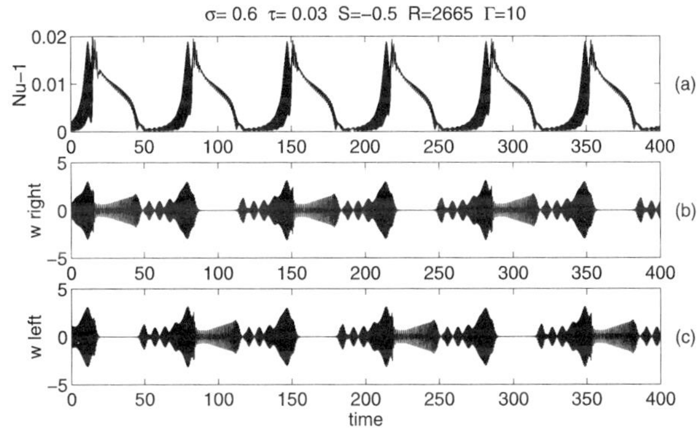

Figure 13 As for fig. 11 but showing periodic bursts at $R = 2665$ ($\epsilon = 8.2 \times 10^{-3}$).

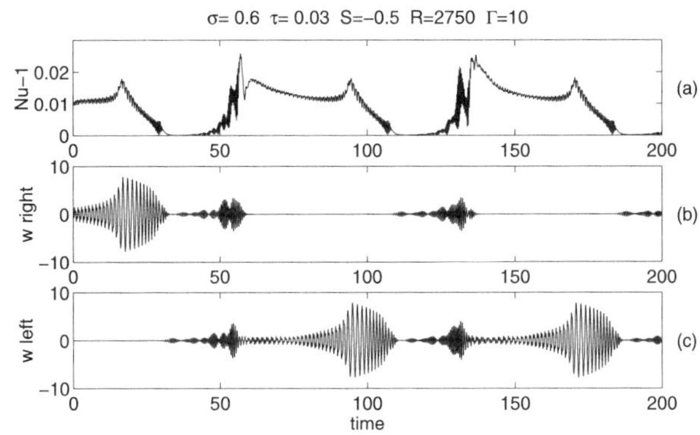

Figure 14 Irregular switching between the fish state and a blinking state at $R = 2750$ ($\epsilon = 0.040$).

that the modulation disappears and (at $R = 2606$) a stable odd parity chevron is recovered. Likewise, fig. 17 shows a chaotic blinking state at $R = 2665$ that is also based on an odd parity chevron and coexists stably with the bursts shown in fig. 13 that are based on an even chevron. This multistability greatly complicates the behavior of the system, and appears to be a consequence of increasing subcriticality of the primary bifurcations with decreasing separation ratio, cf. [10].

5 Origin of the blinking states and repeated transients

In this section we summarize the properties of a simple model system, based on normal form theory for the interaction of two Hopf modes with opposite parity, that accounts for essentially all the properties revealed in our simulations when $\epsilon \ll \Gamma^{-2}$. We do not have a corresponding understanding of the dynamics observed at larger Rayleigh numbers. The model is based on the observation that, at specific aspect ratios Γ_c, the odd and even chevrons bifurcate simultaneously (fig. 2c), albeit with different frequencies (fig. 2d). For nearby aspect ratios they come in in close

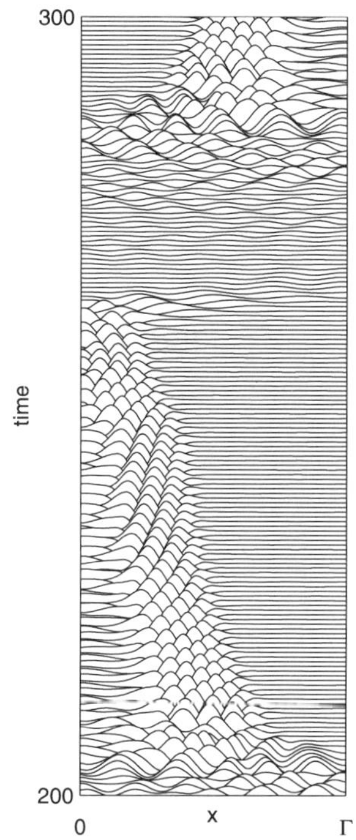

Figure 15 Space-time plot of $\theta(x, 0, t)$ corresponding to the state in fig. 14 with time increasing upwards.

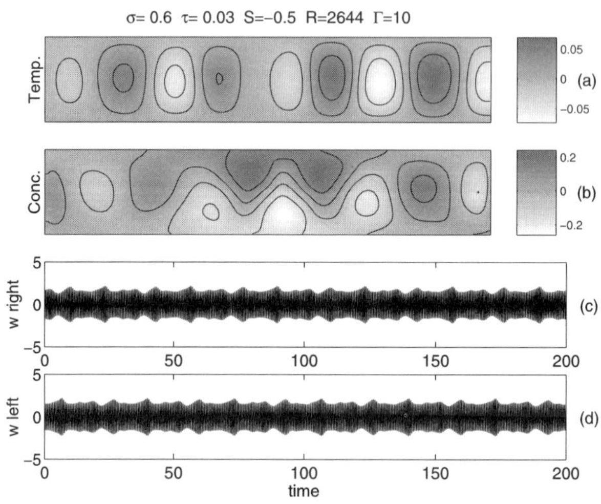

Figure 16 Odd parity chevron state with a superposed small amplitude temporal modulation for $R = 2644$ ($\epsilon = 2.1 \times 10^{-4}$). (a,b) Contours of $\theta(x, z, t)$ and $\Sigma(x, z, t)$ at $t = 200$. (c,d) $w(x = 0.87\Gamma, z = 0, t)$ and $w(x = 0.13\Gamma, z = 0, t)$. This state coexists with the state in fig. 11.

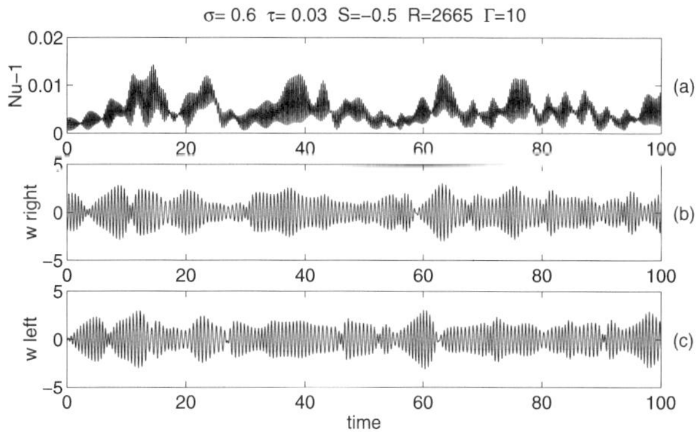

Figure 17 A stable chaotic blinking state at $R = 2665$ ($\epsilon = 8.2 \times 10^{-3}$) based on an odd parity chevron that coexists with the bursts shown in fig. 13.

succession. Under these conditions we may expect the chevrons to interact in the nonlinear regime, and to do so already at small amplitude. We write each of the fields ψ, θ, η in the form

$$\theta(x,z,t) = \mathrm{Re}\{z_+(t)e^{i\omega_+ t} f_+(x,z) + z_-(t)e^{i\omega_- t} f_-(x,z)\} + \dots, \tag{5.1}$$

where $f_\pm(x,z)$ are the (complex) even and odd chevron-like eigenfunctions (Sect. 3), and $z_\pm(t)$ are their amplitudes. Standard normal form theory now yields the following equations for the (real) amplitudes $r_\pm \equiv |z_\pm|$:

$$\dot{r}_+ = (\mu + a_+ r_+^2 + b_+ r_-^2 - r_+^4 + c_+ r_+^2 r_-^2 + d_+ r_-^4) r_+ + \dots, \tag{5.2}$$

$$\dot{r}_- = (\mu - \delta + a_- r_-^2 + b_- r_+^2 - r_-^4 + c_- r_+^2 r_-^2 + d_- r_+^4) r_- + \dots, \tag{5.3}$$

with a pair of decoupled equations for the associated (nonlinear) frequencies $\omega_\pm$. Here the coefficients are all real, and we take $a_\pm > 0$ so that both chevrons bifurcate subcritically, the even ones (given by $(z_+, z_-) = (r_+, 0) \exp i\omega_+ t)$ at $\mu = 0$ and the odd ones (given by $(z_+, z_-) = (0, r_-) \exp i\omega_- t)$ at $\mu = \delta > 0$. It is now a simple matter to show that the two chevrons undergo saddle-node bifurcations at $\mu = -a_\pm^2/4 < 0$, and steady state bifurcations to a mixed parity state (z_+, z_-), $r_+ r_- > 0$, at $\mu = \delta - b_- r_+^2 - d_- r_+^4$ (even chevrons) and $\mu = -b_+ r_-^2 - d_+ r_-^4$ (odd chevrons). The bifurcations to the mixed parity states are to be identified with Hopf (more precisely torus) bifurcations from the chevron states to blinking states. This is because near this bifurcation on the even chevron branch the dynamics takes the form

$$\theta(x,z,t) = \mathrm{Re}\, r_+ e^{i\omega_+ t}\{f_+(x,z) \tag{5.4}$$

$$+ (r_-/r_+) e^{i(\omega_- - \omega_+)t} f_-(x,z)\} + \dots, \tag{5.5}$$

describing an even chevron with a periodically oscillating odd parity contribution. The second term amplifies (reduces) waves in the left half of the container at the same time as it reduces (amplifies) waves in the right half, and shows that the blinking frequency at leading order is simply the beat frequency $\omega_2 = \omega_+ - \omega_-$. We emphasize that these frequencies are the *nonlinear* frequencies, not the onset frequencies predicted by linear theory.

When this bifurcation occurs below the saddle-node bifurcation on the chevron branch the blinking states are initially unstable but acquire stability with increasing amplitude at a *tertiary* Hopf bifurcation, as first pointed out by Langford [18, 36]. This bifurcation introduces a *third* frequency ω_3 into the dynamics of the system. When $a_\pm > 0$ and $\delta > 0$ are small and $b_+ b_- < 0$ this bifurcation is the only one on the branch of the mixed parity states, and occurs when

$$a_+ r_+^2 + a_- r_-^2 - 2(r_+^4 + r_-^4) + (c_+ + c_-) r_+^2 r_-^2 = 0 \,. \tag{5.6}$$

Below we identify the resulting states with the observed repeated transients.

The key transitions involving the one-, two- and three-frequency states are captured by a simplified model system obtained from the above equations by dropping inessential terms. In particular, we drop the term $-r_-^5$ and mimic its effect by taking $a_- < 0$. The resulting model [5]

$$\dot{r}_+ = (\mu + a_+ r_+^2 - r_+^4 - r_-^2) r_+ \tag{5.7}$$

$$\dot{r}_- = (-\nu + a_- r_-^2 + b_- r_+^2) r_- \tag{5.8}$$

with $a_+ > 0$, $b_- > 0$ is the simplest set of equations capable of describing correctly the stability properties of the even chevrons and the mixed parity states observed in the partial differential equations for moderate values of $|S|$. The model removes the primary bifurcation to the odd chevrons but leaves the secondary bifurcations from the even chevrons fundamentally unchanged. In the following we think of μ, ν as proportional to $R - R_c(\Gamma)$, $\Gamma - \Gamma_c(R)$, respectively.

Figure 18 summarizes the properties of the model in the case in which the three-frequency state created from the blinking state branch is stable. This is always the case when $b_- = 1$, $a_- = 0$ and $a_+ > 0$, and hence for sufficiently small negative values of a_- as well. The figure shows the loci of the primary (H_1), secondary (H_2) and tertiary (H_3) Hopf bifurcations, as well as the locus of the saddle-node bifurcations (SN) on the chevron branch. It should be remembered that in the (r_+, r_-) variables only the bifurcation H_3 remains a Hopf bifurcation, with H_1 and H_2 represented by pitchfork bifurcations. In addition the figure shows the curve γ : $\mu = \mu^*(\nu)$ of global bifurcations at which the limit cycle (corresponding to the three-frequency states) created at H_3 disappears by simultaneous collision with small and large amplitude chevron states. The location of this line must be determined numerically. An asymptotic calculation of this curve near the codimension-two point at which H_2 and SN coincide yields the heavy broken line; this line is tangent to γ at the codimension-two point, as it must.

Figure 19a shows the bifurcation diagram obtained by traversing the (μ, ν) plane in fig. 18 along the line $\nu = 1.6$. The figure shows a small interval of subcritical but stable chevrons, followed by a supercritical pitchfork bifurcation to a state with $r_- \neq 0$ that represents a blinking state in the physical variables. In the example shown this bifurcation occurs at $\mu < 0$ so that the first stable state just above onset ($\mu = 0$) is a finite amplitude blinking state. This case is typical of the behavior of the partial differential equations for $S = -0.01$ in appropriate ranges of Γ (not shown). In contrast, when $\nu < a_+ b_-/2 = 1$ the first stable state encountered as μ increases is a periodic state $(r_+(t), r_-(t))$ corresponding to the three-frequency repeated transient state with the frequencies ω_1 and ω_2 filtered out (fig. 19b). This state appears in a global bifurcation at $\mu = \mu^* < 0$ [25], at which $\omega_3 = 0$. Figure 19c shows the time series corresponding to this state when $\mu = 0.02$, $\nu = 0.15$, $a_+ = 2.0$, $a_- = -0.2$, $b_- = 1.0$. For these parameter values

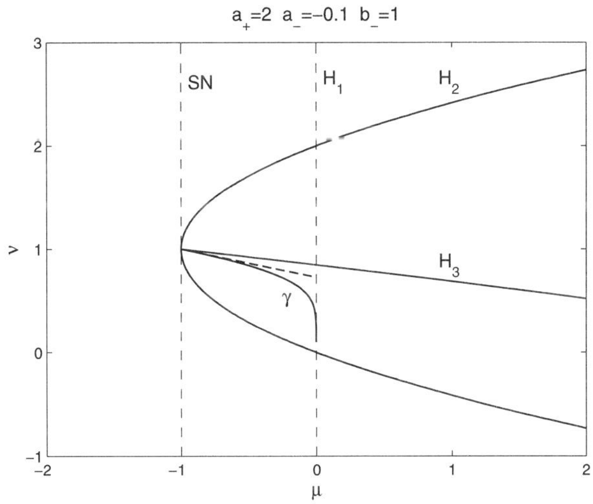

Figure 18 Codimension-one bifurcation surfaces in the (μ, ν) plane for equations (5.7)-(5.8) with $a_+ = 2.0$, $a_- = -0.1$ and $b_- = 1.0$. H_1: primary (Hopf) bifurcation to the chevron state $(r_+, 0)$, SN: saddle-node bifurcation on the chevron branch, H_2: (secondary) Hopf bifurcation to blinking states (r_+, r_-), H_3: (tertiary) Hopf bifurcation from (r_+, r_-) responsible for the appearance of the three-frequency states, and γ: global bifurcation at which these states disappear. The heavy broken line represents the asymptote to γ.

$\mu^* \approx 0$ (open circles in fig. 19b), and fig. 18 shows that this situation persists for a large range of values of ν. In this case the chevron state regrows from very small amplitude, and the resulting oscillation resembles closely the state shown in fig. 4. In particular there is almost no hysteresis between this state and the conduction state, and the system behaves *as if* the primary instability at $\mu = 0$ were directly responsible for generating repeated transients. Observe that during the growth phase of the variable r_+ the variable r_- vanishes, indicating that the growing state is a pure chevron; r_- becomes nonzero only during the collapse phase, indicating that the collapse is triggered by a symmetry-breaking instability of the growing even chevron. The amplitude and the period $2\pi/\omega_3$ of the limit cycle in fig. 19c decrease with increasing μ, with the oscillations disappearing at H_3. As already mentioned we interpret this transition as the transition from the repeated transient state to the (symmetric) periodic blinking state with increasing Rayleigh number seen in fig. 6. For the model parameters this transition is supercritical, indicating absence of hysteresis, as in the figure.

Within the model the repeated transient state $(r_+(t), r_-(t))$ has all the properties of this state observed in the simulations, except for the (apparent) absence of oscillations during the collapse phase. In fact, if the frequencies ω_1 and ω_2 are restored, and the pointwise quantity $[r_+(t) + r_-(t) \cos \omega_2 t] \sin \omega_1 t$, cf. Eq. (5.5), plotted instead of $r_+(t)$ or $r_-(t)$, these oscillations *are* present (fig. 19d), and their amplitude depends on the chevron amplitude r_+ in the manner observed in the simulations. In fact, the time series shown in fig. 19d bears a number of qualitative features, including the pointed overshoot at maximum as the mode r_- begins to

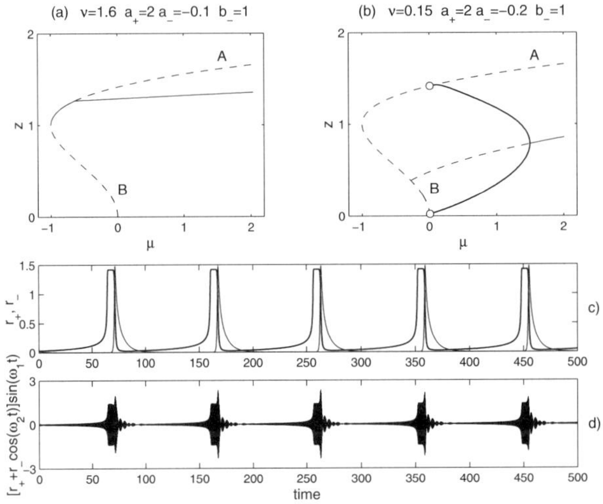

Figure 19 Bifurcation diagrams along (a) the line $\nu = 1.6$ when $a_+ = 2.0$, $a_- = -0.1$, $b_- = 1.0$, and (b) the line $\nu = 0.15$ when $a_+ = 2.0$, $a_- = -0.2$, $b_- = 1.0$. The open circles indicate the global bifurcation; this bifurcation occurs very close to $\mu = 0$ for a large range of values of ν. (c) The time series $r_+(t)$ (thick line) and $r_-(t)$ (thin line) corresponding to case (b) with $\mu = 0.02$. (d) The quantity $[r_+(t) + r_-(t)\cos(\omega_2 t)]\sin(\omega_1 t)$ corresponding to (c) when $\omega_1 = 20.0$, $\omega_2 = 0.8$ for comparison with fig. 4. Note the exponential growth during the chevron phase, followed by an overshoot when the blinking instability sets in, and the ringing down during the subsequent collapse phase.

grow and the "ringing down" due to the fact that the variable r_+ decays more rapidly than r_-, that are documented in experiments as well (fig. 6a of [31]).

Despite its remarkable simplicity the model (5.7)-(5.8) captures completely the two scenarios for generating blinking states identified in the simulations of both ^{3}He-^{4}He and water-ethanol mixtures, and the origin and properties of the repeated transients. Extensions of the model, explored in [5], indicate the possibility that repeated transients may, under appropriate circumstances, be chaotic. In fig. 20 we show an example of such a chaotic repeated transient state. We believe that this state is associated with the global bifurcation in which the repeated transients first appear, cf. fig. 19b. As already noted the frequency ω_3 decreases to zero as $\mu \downarrow \mu^* < 0$, i.e., as $\epsilon \downarrow \epsilon^* < 0$ in the partial differential equations. As this occurs the three-frequency states approach simultaneously the unstable large and small amplitude chevron states, hereafter A and B, respectively (see fig. 19). The character of the repeated transient when $\epsilon \approx \epsilon^*$ is determined by the leading eigenvalues of A and B in the $(r_+, 0)$ direction, hereafter $-\lambda_A < 0$ and $\lambda_B > 0$, and the leading eigenvalues in the $(0, r_-)$ direction. If the latter are real, $\alpha_A > 0$ and $-\alpha_B < 0$, say, and $\rho \equiv \alpha_B\lambda_A/\alpha_A\lambda_B > 1$, the repeated transients will remain periodic and stable all the way to ϵ^*, where the period T_3 diverges and the global bifurcation takes place [5]. In contrast, when $0 < \rho < 1$, the periodic oscillations necessarily lose stability before the global bifurcation at ϵ^*. Similar results obtain in the case where the leading stable symmetry-breaking eigenvalue at B is complex, viz. $-\alpha_B + i\omega_B$, $\alpha_B > 0$, as suggested by the simulations. In this case stable periodic oscillations

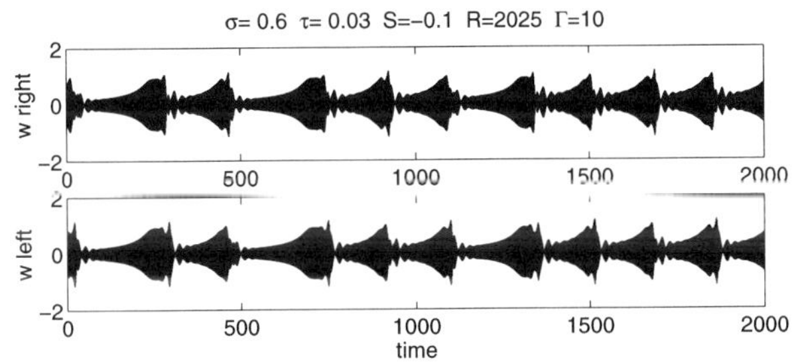

Figure 20 A chaotic repeated transient in a ^{3}He-^{4}He mixture with stress-free and fixed temperature boundary conditions at $x = 0, \Gamma$ and $R = 2025$, $S = -0.1$, $\sigma = 0.6$, $\tau = 0.03$, $\Gamma = 10$.

will persist down to ϵ^* if $\rho > 1$, but if $0 < \rho < 1$ complex dynamics of Shil'nikov type will be present. In fact the leading unstable eigenvalues α_A and λ_B are also expected to be complex, since in the partial differential equations the bifurcations at H_1 and H_2 are both Hopf bifurcations.

When λ_B is real a trajectory escaping from B describes an exponentially growing chevron state. This growth phase, including the states A and B, is clearly visible in the time series in fig. 4. When the growing chevron reaches the vicinity of A it becomes unstable to symmetry-breaking oscillations which take it back near B. This is the collapse phase of the repeated transient state (compare fig. 19d with fig. 4). The frequency of the decaying oscillations observed in the time series in figs. 19d is given by ω_B. This frequency will in general be of the same order as the blinking frequency associated with the branch of blinking states when these bifurcate from the small amplitude chevron B, but quite different from (and in general larger than) the blinking frequency of the *stable* blinking states beyond H_3, cf. [31]. This observation explains the coincidence of the period of the blinking states and of the oscillations during the collapse phase of the repeated transient noted by Kolodner. Note also that since α_B decreases as ϵ decreases (it passes through zero at H_2, i.e., at $\epsilon = \epsilon_2$) the collapse becomes slower and slower, cf. fig. 4, although the collapse rate is still finite when the three-frequency states disappear in the global bifurcation at ϵ^* (since $\epsilon_2 < \epsilon^* < 0$) and the system makes a hysteretic transition back to the conduction state. The fact that α_B decreases with ϵ makes it likely that the Shil'nikov condition $0 < \rho < 1$ holds at ϵ^*, resulting in *chaotic* repeated transients prior to their disappearance. However, despite this suggestion we have only succeeded in locating such chaotic repeated transients with *stress-free* boundary conditions (fig. 20).

6 Localized states and convectons

As R is increased further we encounter new structures: stable localized states. Figure 21 shows an example computed with $S = -0.5$: a stable wall-attached traveling wave present at $R = 2900$. A wave of this type evolves from the fish state when the localized low frequency state comes to rest against a sidewall, but does not collapse. The resulting state should be interpreted in terms of a stationary *front* separating an exponentially small wave throughout most of the domain from

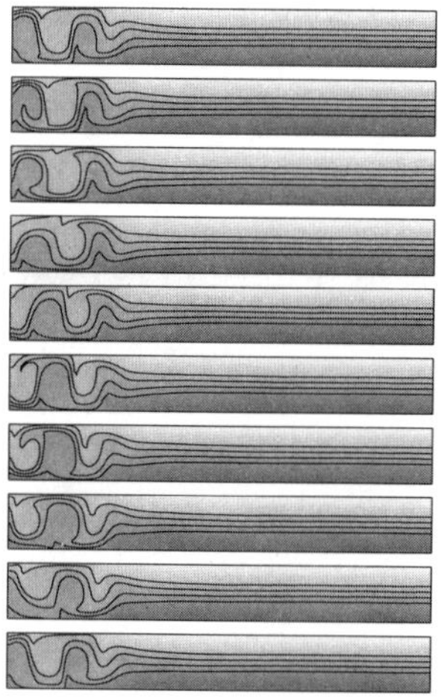

Figure 21 A left-traveling wave confined to the left sidewall when $R = 2900$ ($\epsilon = 0.097$), and $S = -0.5$, $\sigma = 0.6$, $\tau = 0.03$, $\Gamma = 10$, shown in terms of $C(x, z)$, with time increasing upwards.

a finite amplitude wavetrain next to the left sidewall. Thus fig. 21 represents a stable, dynamically localized wave, in contrast to the unstable dynamically localized waves that briefly form for smaller values of R (but satisfying $\epsilon \sim \Gamma^{-2}$). Note, in particular, the absence of a right-traveling wave in any part of the container except perhaps right next to the *left* sidewall where a right-traveling component is required to set up a standing oscillation. Similar states have been observed in water-ethanol experiments [19, 39, 30] and simulations [50, 42]. States of this type are described well by a *single* complex Ginzburg-Landau equation with a drift, and become possible only once $|\epsilon| \gg \Gamma^{-2}$.

With increasing R the strong nonlinear dispersion forces the wave frequency towards zero; once this occurs the resulting non-oscillatory state begins to expand towards the right by adding (steady) rolls and thereby expelling the lateral concentration gradient set up by the confined traveling wave. The process of adding rolls terminates once this lateral gradient is sufficiently strong, and results in the formation of a spatially confined but *steady* state (see fig. 22) we call a *convecton* [7]. Such confined steady states have also been found for other values of R (fig. 23), indicating that the confined states created by this process are in general non-unique. All of these states are numerically stable. Continuation methods indicate that these confined steady states retain their character with varying Rayleigh number, and lie on disconnected solution branches (fig. 24). In particular, the localized traveling wave in fig. 21 is located on a *different* branch of solutions than the states shown in figs. 22 and 23.

To understand the origin of these remarkable states we have integrated equations (2.1)-(2.4) with *periodic* boundary conditions in the horizontal and larger (spatial period) Γ. In fig. 25 we show two *coexisting* convectons computed with periodic boundary conditions when $R = 2800$, $\Gamma = 20$. Figure 25a shows an odd

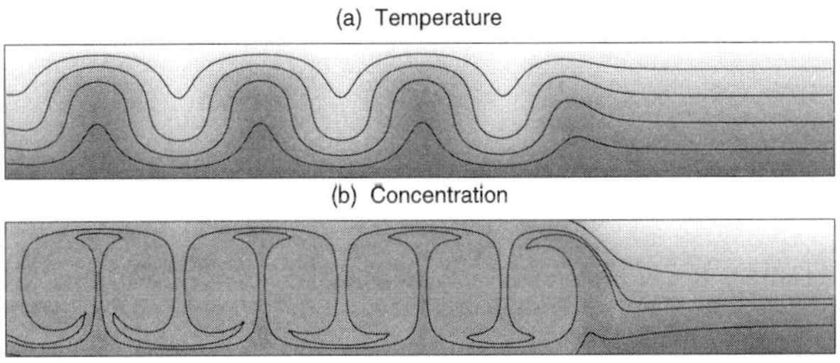

Figure 22 A stable convecton at $R = 3050$ ($\epsilon = 0.154$) in terms of (a) $T(x,z) = 0.5 - z + \theta(x,z)$, (b) $C(x,z)$, when $S = -0.5$, $\sigma = 0.6$, $\tau = 0.03$, $\Gamma = 10$.

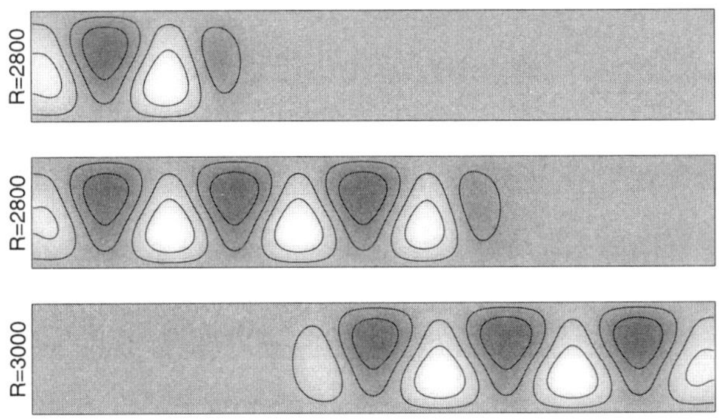

Figure 23 Additional stable convectons when $S = -0.5$, $\sigma = 0.6$, $\tau = 0.03$, $\Gamma = 10$, shown in terms of $\theta(x,z)$. The corresponding value of R is indicated at the left of each panel.

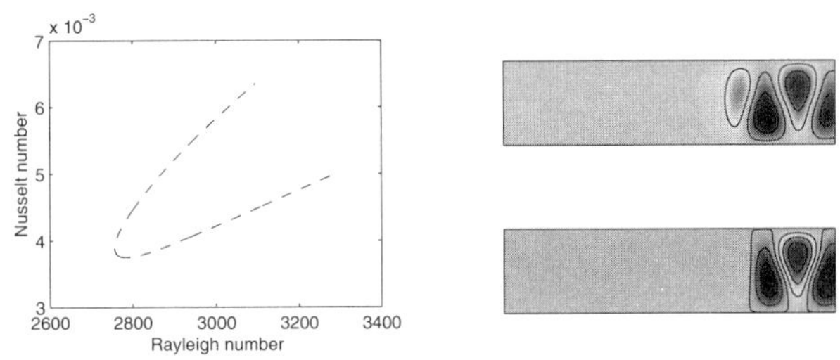

Figure 24 A convecton branch corresponding to the state shown in fig. 23a. The panels on the right show the structure of the convecton in terms of $\theta(x,z)$ at the two ends of the branch. The upper convecton is stable between the turning-point and $R \approx 2800$; the lower is unstable.

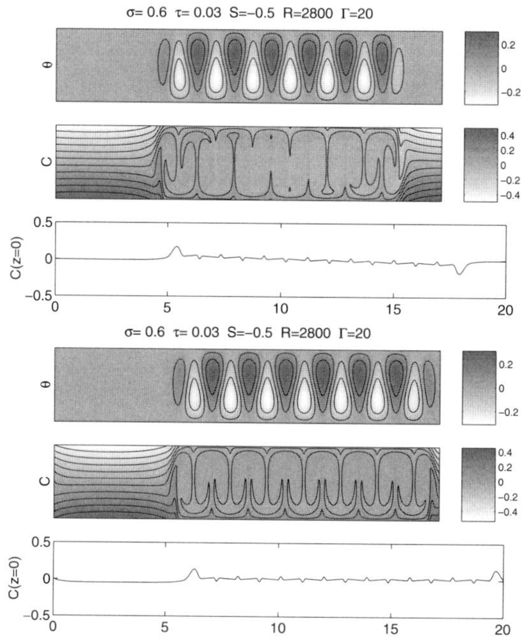

Figure 25 (a) Odd and (b) even parity convectons at $R = 2800$ ($\epsilon = 0.1681$) in a ^{3}He-^{4}He mixture with $S = -0.5$, $\sigma = 0.6$, $\tau = 0.03$ and spatial period $\Gamma = 20$. In each figure the top panel shows contours of constant θ, the middle panel shows contours of constant concentration C and the bottom panel shows $C(x, z = 0)$.

parity convecton, while (b) is of even parity; both are time-independent and numerically stable. Each state is described in terms of the contours of the temperature fluctuation $\theta(x,z)$ and of the associated concentration $C(x,z)$. In addition the lowest panel shows the midplane concentration. Both states consist of steady rolls, separated by two fronts from a convection-free region. Since concentration gradients are expelled from regions of closed streamlines the rolls are well mixed, with large concentration gradients along the top and bottom boundaries. The resulting step-like profile in $C(x, z = 0)$ (fig. 25a) is typical of this type of expulsion process [27], with dips in $C(x, z = 0)$ associated with downflow, and bumps with upflow. The inhomogeneity in the roll amplitude at the boundary of the convecton amplifies this effect and produces larger concentration anomalies at its edges; these anomalies in turn 'trap' the convecton, much as discussed by Riecke in the context of self-trapping of wave packets [44]. We may think of these anomalies as the location of the bounding fronts. In contrast the temperature boundary layers are substantially weaker, as expected in a system with $\tau \ll 1$, and do not contribute to self-trapping. Note that despite the fact that it is not left-right symmetric the odd convecton is in fact *stationary*. This is a consequence of the fact that it is symmetric under the reflection κR_x [26]. Evidently, if the fixed temperature boundary condition at the top (or the bottom) were changed into a Newton's law of cooling the odd parity convecton would drift at a constant speed, while the even one would remain at rest.

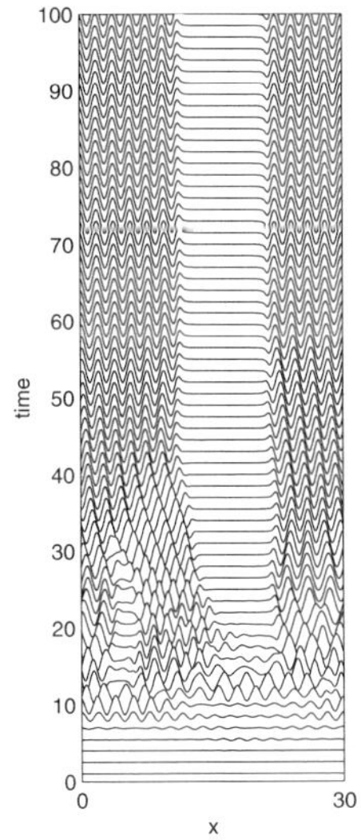

Figure 26 The formation of an odd parity convecton when $R = 2800$, $\Gamma = 30$ starting from small amplitude initial conditions, in the form of a space-time plot of $\theta(x, z = 0, t)$ with time increasing upward.

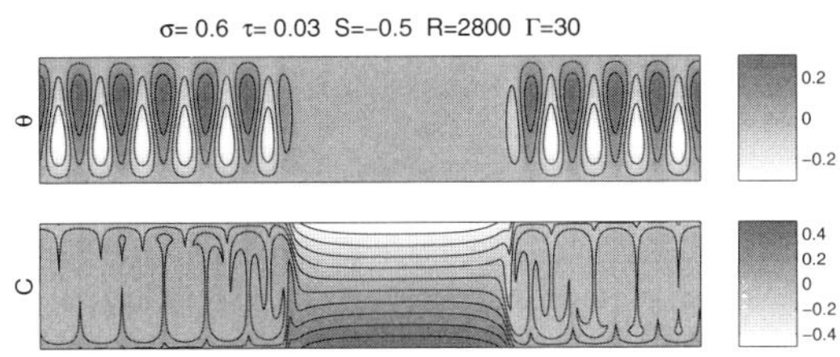

Figure 27 An odd parity convecton formed in the process shown in fig. 26. The resulting convecton contains a larger number of rolls than the one shown in fig. 25a.

Figure 26 shows the evolution of a small amplitude initial perturbation into a convecton in the form of a space-time plot when $R = 2800$, $\Gamma = 30$ with time increasing upwards in units of the thermal diffusion time in the vertical. We see that the initial perturbation disperses creating a region of lower amplitude larger wavelength oscillations at its original location; the resulting longer wavelength oscillations are more effective at expelling concentration gradients, and hence suppressing their own growth, and creating a void. This void 'pushes' the remaining rolls into an incipient convecton and compresses the rolls within it. In the stationary state the convecton consists of stationary rolls of uniform amplitude, except for two 'front'-like structures at either end that separate it from the void. Figure 27 shows the final state.

As R increases we find that the number of rolls within the convecton slowly increases. The transitions leading to the addition of a roll are hysteretic. However,

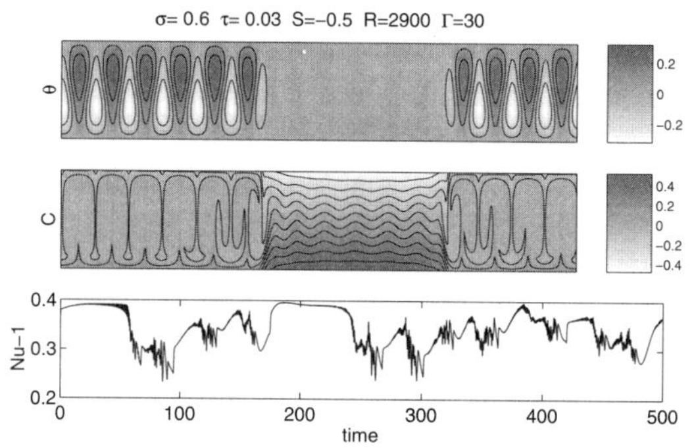

Figure 28 Confined convecton-like state maintained by traveling waves in the void region when $R = 2900$, $\Gamma = 30$ and $t = 500$. The third panel shows the fluctuations in the Nusselt number arising from fluctuations in the number of rolls in the structure.

once R reaches $R = 2900$ the behavior changes dramatically, with the void region now filled with traveling waves (fig. 28). These propagate outward from the void center, growing in amplitude as they travel, and are responsible for the irregular oscillations of the interface between the convecton and void; these are in turn responsible for the fluctuations in the number of rolls in the convecton, as seen in the Nusselt number Nu(t) (fig. 28).

In contrast when R is decreased the number of rolls within the convecton gradually decreases. Figure 29 shows the gradual decrease in the Nusselt number when R is reduced from $R = 2800$ to (a) $R = 2750$ and (b) $R = 2700$. In (a) the convecton simply becomes more compact while in (b) each of the barely visible oscillations in Nu(t) corresponds to the expulsion of a pair of rolls. Figure 30 shows the space-time plot corresponding to the destruction of the initial convecton in fig. 29b. These results are broadly consistent with existing experiments on water-ethanol mixtures [32, 33]. However, if R is sufficiently small (as in fig. 29b) that a convecton consisting of two rolls is unstable the expulsion process does not stop, and the roll amplitude falls to a small value before a *new* convecton regrows much as in fig. 26. Figure 29b shows the remarkable relaxation oscillations that result, and indicates that the convectons created in this nucleation process may have different numbers of rolls.

Localized steady states consisting of a finite number of roll-like structures generally occur as a consequence of spatial locking of the fronts bounding the state and the periodic structure within the state [43]. In this picture the localized states are associated with homoclinic connections to the flat state. For variational systems with reflection symmetry in one spatial dimension the process generating such states is well understood [22]: the states exist in a finite parameter interval and are described by a pair of intertwined curves of homoclinics (one for odd states, the other for even states) that oscillate back and forth as they approach the spatially periodic perfect roll state; with each turn the localized state acquires a pair of additional 'rolls'. Thus there is a parameter interval with an infinite number of coexisting localized states of both parities; these appear in a sequence of saddle-node bifurcations involving stable and unstable solutions. States of this type have been

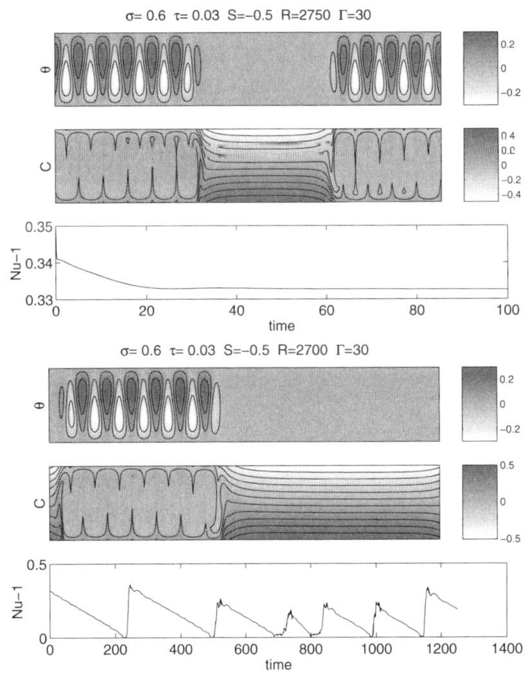

Figure 29 The evolution of the odd parity convecton in fig. 27 when R is decreased from $R = 2800$ to (a) $R = 2750$, (b) $R = 2700$, for $\Gamma = 30$. The lower panels show the evolution of the Nusselt number; the upper panels show the state at the last instant.

seen in the Swift-Hohenberg equation with both quadratic and cubic terms [20]. Moreover, when the bifurcation parameter ϵ is reduced below the lowest value at which a stationary front is present the structured state invades the uniform state in 'jumps' of one wavelength of the structure. Each of these states can in principle be restabilized by increasing ϵ again. This picture extends to nonvariational systems, like binary fluid convection, as well, and indeed the space-time plots in figs. 26 and 30 resemble those computed in [11, 12].

The above theory, however, applies only when the homogeneous and periodic states are both stable. We believe that the existence of convectons when $R > R_c$ is a consequence of two additional physical mechanisms. The expulsion of concentration gradients increases locally the value of ϵ by depressing R_c, while in the void regions the concentration difference between the bottom and top is unchanged. In these regions $\epsilon > 0$ but no instability occurs because the convectons serve as absorbing boundaries. These raise the instability threshold sufficiently to stabilize the conduction state in the void region: $0 < \epsilon < \epsilon_c^{void}$. For example, the void region associated with the convecton in fig. 27 is of length $\Gamma_{void} \approx 10$; the critical Rayleigh number for the void with no-flux no-slip lateral boundary conditions is then $R_c^{void} \approx 2643$ and this number can be increased by reducing the effective reflection coefficient at the lateral boundaries. We believe therefore that the void regions correspond to a locally stable conduction state; once $R > R_c^{void}$ the void region fills with traveling waves (fig. 28), much as observed in water-ethanol mixtures [32].

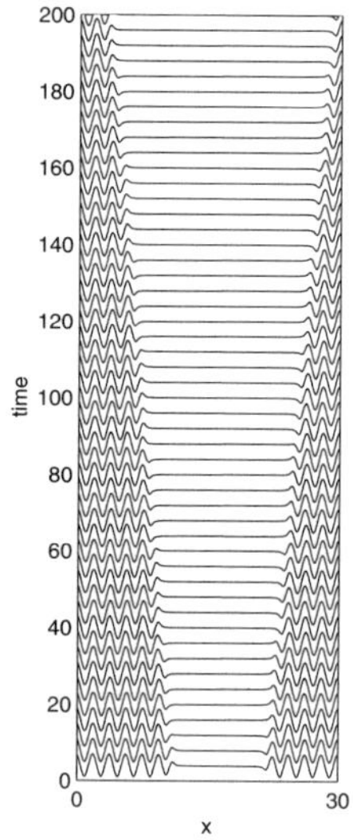

Figure 30 The destruction of the odd parity convecton when R is decreased from $R = 2800$ to $R = 2700$ (fig. 29b) in the form of a space-time plot of $\theta(x, z = 0, t)$. Both fronts recede at the same rate preserving the parity of the state. The convecton eventually reforms (fig. 29b) and the slow erosion starts over again.

7 Discussion

In this article we have described the results of direct numerical simulations of oscillatory convection in binary mixtures in two-dimensional moderately large domains with experimentally relevant boundary conditions, exploring in detail the parameter values characteristic of ^{3}He-^{4}He mixtures. We have chosen ^{3}He-^{4}He parameters partly because of a number of experiments on this system in the 1980's [46, 1, 47, 8] and partly to make predictions for future experiments now that flow visualization at cryogenic temperatures has become possible [49]. We have also used smaller domains in order to reduce the length of the immensely long transients encountered in our simulations of water-ethanol mixtures [5]. As a result we have been able to explore a larger range of values of the Rayleigh number, reaching values at which the low-dimensional description of this system valid near threshold begins to breakdown. We have argued that this transition occurs once $|\epsilon| \sim \Gamma^{-2}$ and provided numerical evidence in support of this claim. Among the new phenomena that occur in this regime are the *fish states* originally observed by Kolodner et al. in water-ethanol mixtures [34] and various types of dynamically localized traveling wave states also seen in experiments [19, 39, 30].

For $|\epsilon| \ll \Gamma^{-2}$ the dynamics of the system can be understood in detail using a finite-dimensional description based on the interaction of adjacent pure parity chevron states whose structure is known from linear theory. These states interact strongly at small amplitude in the vicinity of mode interaction points, i.e., specific values of the aspect ratio Γ at which both modes bifurcate from the conduction state simultaneously (see fig. 2c). These mode interaction points correspond to a double Hopf bifurcation with generically nonresonant frequencies. Unfolding of the normal form for this bifurcation to take into account the splitting of the mode interaction leads to a pair of simple equations for the amplitudes of the two competing modes. It is a remarkable fact that these equations describe all the qualitative behavior of the system for $|\epsilon| \ll \Gamma^{-2}$, including the blinking and repeated transient states and the transitions between them, provided only that one takes into account the fact that the primary instability to the chevron state is generally subcritical. In particular, the model constructed in Section 5 shows that the repeated transients are a three-frequency state, and explains why it is so often the first nontrivial state observed in the simulations once the conduction state loses stability, and why it often appears without observable hysteresis. The model also shows that with increasing Rayleigh number the lowest of the three frequencies in the repeated transient state disappears in a tertiary Hopf bifurcation, leaving behind a stable large amplitude blinking state. In the model, as in the simulations, this bifurcation is supercritical and hence is nonhysteretic. In contrast, the corresponding bifurcation in water-ethanol mixtures was found to be always subcritical and hence hysteretic [5]. Finally, the model also suggests the possibility of chaotic repeated transients, and we were able to find such states in ^{3}He-^{4}He mixtures, although only with stress-free boundary conditions. In principle the coefficients in the model can be calculated from equations (2.1)-(2.6) in terms of the physical parameters of the system. This is a major undertaking, however, and is beyond the scope of this article.

The model (5.7)-(5.8) omits other important transitions that may occur. In particular this is the transition from the chevron states to asymmetric but single-frequency chevron-like states called TW$'$ in [14, 15]. In large domains these states come to resemble spatially uniform traveling waves at large enough ϵ. Such states do not appear to be present when $|\epsilon| \ll \Gamma^{-2}$, a fact reflected in the absence of frequency locking in the model, although they are found in the unfolding of certain codimension-three singularities, cf. Sect. 3.

The fish state (fig. 8) is the first manifestation of the transition to a large number of degrees of freedom ($\epsilon \sim \Gamma^{-2}$), the fish phase corresponding to temporary but dynamic confinement of the traveling wave state. The hallmark of dynamic confinement is the essentially complete absence of counterpropagating waves. This is seen not only in the wall-attached state in fig. 21 but also in the space-time plots in figs. 9 and 15. In a dynamically confined state the finite amplitude traveling waves are separated from the convection-free region (i.e., a region of exponentially small waves) by a sharp front of $O(\epsilon^{-1/2})$ width; the transition between this regime and the low-dimensional regime dominated by chevron-like states occurs when the width of the front becomes comparable to the size of the domain, i.e., $|\epsilon|\Gamma^2 \sim 1$. The latter regime is dominated by sidewalls and any localization can be at most kinematic [15]. In the kinematic regime the amplitude of the counterpropagating wave is never zero, and a kinematically localized state near the left sidewall always shows a small amplitude right-traveling wave near the right sidewall, i.e., such a localized state is nothing but a strongly asymmetric chevron, cf. [15, 13]. Our simulations are broadly consistent with these ideas. For example, in fig. 8 $\Gamma = 10$ and $\epsilon = 2.35 \times 10^{-2}$ so that $|\epsilon|\Gamma^2 = 2.35$, while in [34], where the fish state was first observed, $\Gamma = 16.75$ and $\epsilon = 0.0103$ giving $|\epsilon|\Gamma^2 = 2.89$. Both of these examples represent marginal cases just outside of the low-dimensional description. The experiment [24] on water-ethanol mixtures is further in the high-dimensional regime: here $\Gamma = 40.6$ and $\epsilon = 0.0094$ giving $|\epsilon|\Gamma^2 = 15.49$. Indeed [24] argue that for sufficiently large ϵ the dynamics of the system resemble more and more the collapse events described by a single subcritical Ginzburg-Landau equation on an *unbounded* domain. In fact the experiments [24] suggest that this is so once $|\epsilon|\Gamma^2 \sim 10$, while [41], performed for slightly larger $|S|$ and subcritical values of ϵ, suggests the requirement $|\epsilon|\Gamma^2 \sim 5$. Our simulations indicate that the location of the transition to high-dimensional behavior depends quite sensitively on the value of $|S|$, although in all cases $|\epsilon|\Gamma^2$ remains of order one. For example, our results for $S = -0.021$ [3, 4] yield $|\epsilon|\Gamma^2 \sim 0.3$; this value is *identical* to that for which [46] find 'intermittent' oscillations in the Nusselt number in a ^{3}He-^{4}He experiment ($\epsilon = 0.3 \times 10^{-3}$, $\Gamma = 34$) although other explanations for this observation have been suggested [38]. For $S = -0.5$ the transition to the fish state and subsequent transitions probably all take place at negative values of ϵ, not considered here.

The temporarily localized traveling wave states shown in fig. 10 resemble in structure the so-called LTW states present with laterally periodic boundary conditions and water-ethanol parameters [2]. However, in our case these states have a lower frequency than the corresponding chevrons, in contrast to the LTW which have a substantially higher frequency than the spatially uniform TW at the same parameters [2, 23]. Of course our localized states are unstable and may not be related to the LTW which are unaffected by lateral walls. In addition our states are present at substantially positive values of ϵ. It is noteworthy that with our boundary conditions we have been unable to find stable finite amplitude traveling waves that fill the entire domain. All our solutions always eventually break up into the type of states described in Section 4. Although there is no evidence that ^{3}He-^{4}He mixtures support spatially homogeneous traveling waves such waves have been observed in rectangular containers filled with water-alcohol mixtures. This unexpected difference may be related to the fact that traveling waves in ^{3}He-^{4}He mixtures are typically more subcritical than in water-ethanol mixtures [10], while the opposite is the case for standing waves (i.e., chevrons, in finite domains). To our knowledge no convectons have been observed in any binary mixture experiment.

Acknowledgments: We are grateful to I. Mercader and M. Net for assistance and to A. Champneys for a helpful discussion.

References

[1] Ahlers, G. and Rehberg, I. *Convection in a binary mixture heated from below*, Phys. Rev. Lett. **56** (1986), 1373–1376.

[2] Barten,W., Lücke, M., Kamps, M. and Schmitz, R. *Convection in binary fluid mixtures. II. Localized traveling waves*, Phys. Rev. E **51** (1995), 5662–5680.

[3] Batiste, O. and Knobloch, E. *Oscillatory binary fluid convection in finite containers*, in Perspectives and Problems in Nonlinear Science, E. Kaplan, J. Marsden, and K. Sreenivasan (eds.), Springer-Verlag, Berlin, 2003, pp. 91–144.

[4] Batiste, O. and Knobloch, E. *Simulations of oscillatory convection in ^{3}He-^{4}He mixtures in moderate aspect ratio containers*, Phys. Fluids **17** (2005), 064102 (14pp).

[5] Batiste, O., Knobloch, E., Mercader, I. and Net, M. *Simulations of oscillatory binary fluid convection in large aspect ratio containers*, Phys. Rev. E **65** (2001), 016303 (19pp).

[6] Batiste, O., Mercader, I., Net, M. and Knobloch, E. *Onset of oscillatory binary fluid convection in finite containers*, Phys. Rev. E **59** (1999), 6730–6741.

[7] Blanchflower, S. *Magnetohydrodynamic convectons*, Phys. Lett. A **261** (1999), 74–81.

[8] Bloodworth, T. J., Ardron, M. R., Bhattacharjee, J. K., Lucas, P. G. J. and Stein, N. D. *Convection in ^{3}He-^{4}He mixtures with large negative separation ratios*, Nonlinearity **3** (1990), 981–995.

[9] Caldwell, D. R. *Experimental studies on the onset of thermohaline convection*, J. Fluid Mech. **64** (1974), 347–367.
[10] Clune, T. and Knobloch, E. *Mean flow suppression by endwalls in oscillatory binary fluid convection*, Physica D **61** (1992), 106–112.
[11] Coullet, P., Riera, C. and Tresser, C. *Stable static localized structures in one dimension*, Phys. Rev. Lett. **84** (2000), 3069–3072.
[12] Coullet, P., Riera, C. and Tresser, C. *Qualitative theory of stable stationary localized structures in one dimension*, Prog. Theor. Phys. Suppl. **139** (2000), 46–58.
[13] Croquette, V. and Williams, H. *Nonlinear competition between waves on convection rolls*, Phys. Rev. A **39** (1989), 2765–2768.
[14] Dangelmayr, G. and Knobloch, E. *Hopf bifurcation with broken circular symmetry*, Nonlinearity **4** (1991), 399–427.
[15] Dangelmayr, G., Knobloch, E. and Wegelin, M. *Travelling wave convection in finite containers*, Europhys. Lett. **16** (1991), 723–729.
[16] Daniels, P. G. *Asymptotic sidewall effects in rotating Bénard convection*, Z. Angew. Math. Phys. **28** (1977), 577–584.
[17] Deane, A. E., Knobloch, E. and Toomre, J. *Travelling waves in large-aspect-ratio thermosolutal convection*, Phys. Rev. A **37** (1988), 1817–1820.
[18] Guckenheimer, J. *On a codimension-two bifurcation*, in Dynamical Systems and Turbulence, Lecture Notes in Mathematics **898**, D. A. Rand and L. S. Young (eds.), Springer-Verlag, New York, 1981, pp. 99–142.
[19] Heinrichs, R., Ahlers, G. and Cannell, D. S. *Traveling waves and spatial variation in the convection of a binary mixture*, Phys. Rev. A **35** (1987), 2761–2764.
[20] Hilali, M'F., Métens, S., Borckmans, P. and Dewel, G. *Pattern selection in the generalized Swift-Hohenberg model*, Phys. Rev. E **51** (1995), 2046–2052.
[21] Hirschberg, P. and Knobloch, E. *Mode interactions in large aspect ratio convection*, J. Nonlinear Sci. **7** (1997), 537–556.
[22] Hunt, G. W., Peletier, M. A., Champneys, A. R., Woods, P. D., Wadee, M. A., Budd, C. J. and Lord, G. J. *Cellular buckling in long structures*, Nonlin. Dyn. **21** (2000), 3–29 and references therein.
[23] Jung, D. and Lücke, M. *Localized waves without the existence of extended waves: oscillatory convection of binary mixtures with strong Soret effect*, Phys. Rev. Lett. **89** (2002), 054502 (4pp).
[24] Kaplan, E., Kuznetsov, E. and Steinberg, V. *Burst and collapse in traveling-wave convection of a binary fluid*, Phys. Rev. E **50** (1994), 3712–3722.
[25] Kirk, V. and Knobloch, E. *A remark on heteroclinic bifurcations near steady state/pitchfork bifurcations*, Int. J. Bif. Chaos **14** (2004), 3855–3869.
[26] Knobloch, E. *Symmetry and instability in rotating hydrodynamic and magnetohydrodynamic flows*, Phys. Fluids **8** (1996), 1446–1454.
[27] Knobloch, E. and Merryfield, W. J. *Enhancement of diffusive transport in oscillatory flows*, Astrophys. J. **401** (1992), 196–205.
[28] Knobloch, E., Proctor, M. R. E. and Weiss, N. O. *Finite-dimensional description of doubly diffusive convection*, in Turbulence in Fluid Flows: A Dynamical Systems Approach, IMA Volumes in Mathematics and its Applications **55**, G. R. Sell, C. Foias, and R. Temam (eds.), Springer-Verlag, New York, 1993, pp. 59–72.
[29] Kolodner, P. *Neutrally stable fronts of slow convective traveling waves*, Phys. Rev. A **42** (1990), 2475–2478.
[30] Kolodner, P. *Stable and unstable pulses of traveling-wave convection*, Phys. Rev. A **43** (1991), 2827–2832.
[31] Kolodner, P. *Repeated transients of weakly nonlinear traveling-wave convection*, Phys. Rev. E **47** (1993), 1038–1048.
[32] Kolodner, P. *Coexisting traveling waves and steady rolls in binary-fluid convection*, Phys. Rev. E **48** (1993), R665–R668.
[33] Kolodner, P., Slimani, S., Aubry, N. and Lima, R. *Characterization of dispersive chaos and related states of binary fluid convection*, Physica D **85** (1995), 165–224.
[34] Kolodner, P., Surko, C. M. and Williams, H. *Dynamics of traveling waves near the onset of convection in binary fluid mixtures*, Physica D **37** (1989), 319–333.

[35] Landsberg, A. S. and Knobloch, E. *Oscillatory bifurcation with broken translation symmetry*, Phys. Rev. E **53** (1996), 3579–3600.

[36] Langford, W. F. *A review of interactions of Hopf and steady-state bifurcations*, Nonlinear Dynamics and Turbulence, G. I. Barenblatt, G. Iooss and D. D. Joseph (eds), Pitman, Boston, 1983, pp. 215–237.

[37] Martel, C. and Vega, J. M. *Dynamics of a hyperbolic system that applies at the onset of the oscillatory instability*, Nonlinearity **11** (1998), 105–142.

[38] Moehlis, J. and Knobloch, E. *Forced symmetry breaking as a mechanism for bursting*, Phys. Rev. Lett. **80** (1998), 5329–5332.

[39] Moses, E., Fineberg, J. and Steinberg, V. *Multistability and confined traveling-wave patterns in a convecting binary mixture*, Phys. Rev. A **35** (1987), 2757–2760.

[40] Nagata, W. *Convection in a layer with sidewalls – bifurcation with reflection symmetries*, Z. Angew. Math. Phys. **41** (1990), 812–828.

[41] Niemela, J. J., Ahlers, G. and Cannell, D. S. *Localized traveling-wave states in binary-fluid convection*, Phys. Rev. Lett. **64** (1990), 1365–1368.

[42] Ning, L., Harada, Y. and Yahata, H. *Modulated traveling waves in binary fluid convection in an intermediate-aspect-ratio rectangular cell*, Prog. Theor. Phys. **97** (1997), 831–848.

[43] Pomeau, Y. *Front motion, metastability and subcritical bifurcations in hydrodynamics*, Physica D **23** (1986), 3–11.

[44] Riecke, H. *Self-trapping of traveling-wave pulses in binary mixture convection*, Phys. Rev. Lett. **68** (1992), 301–304.

[45] Steinberg, V., Fineberg, J., Moses, E. and Rehberg, I. *Pattern selection and transition to turbulence in propagating waves*, Physica D **37** (1989), 359–383.

[46] Sullivan, T. S. and Ahlers, G. *Nonperiodic time dependence at the onset of convection in a binary liquid mixture*, Phys. Rev. A **38** (1988), 3143–3146.

[47] Sullivan, T. S. and Ahlers, G. *Hopf bifurcation to convection near the codimension-two point in a ^{3}He-^{4}He mixture*, Phys. Rev. Lett. **61** (1988), 78–81.

[48] van Gils, S. A., Krupa, M. and Langford, W. F. *Hopf bifurcation with non-semisimple 1:1 resonance*, Nonlinearity **3** (1990), 825–850.

[49] Woodcraft, A. L., Lucas, P. G. J., Matley, R. G. and Wong, W. Y. T. *Visualisation of convective flow patterns in liquid helium*, J. Low Temp. Phys. **114** (1999), 109–134.

[50] Yahata, H. *Travelling convection rolls in a binary fluid mixture*, Prog. Theor. Phys. **85** (1991), 933–937.

Fields Institute Communications
Volume **49**, 2006

Mixed Mode Oscillations Due to the Generalized Canard Phenomenon

Morten Brøns
Department of Mathematics
Technical University of Denmark
DK-2800 KGS.Lyngby, Denmark
M.Brons@mat.dtu.dk

Martin Krupa
Department of Mathematical Sciences
New Mexico State University
Las Cruces, NM 88003 USA
mkrupa@nmsu.edu

Martin Wechselberger
School of Mathematics & Statistics
University of Sydney
NSW 2006 Australia
M.Wechselberger@maths.usyd.edu.au

Abstract. Mixed mode oscillations combine features of small oscillations and large oscillations of relaxation type. We describe a mechanism for mixed mode oscillations based on the presence of canard solutions, which are trajectories passing from a stable to an unstable slow manifold. An important ingredient of this mechanism are singularities known as folded nodes. The main focus of this article is to show how the local dynamics near a folded node can combine with global features, leading to mixed mode oscillations. We review and extend the results of [28] on the dynamics near a folded node and state some results on mixed mode periodic orbits with Farey sequences of the form 1^s. We also show how to generalize the context of one fast variable to an arbitrary number of fast variables.

2000 *Mathematics Subject Classification.* Primary 34C26; Secondary 34C30, 34D15, 37D10, 37N25.

M. Krupa was supported in part by the National Science Foundation grant No. 0406608 and by the Otto Mønsted Foundation.

Martin Wechselberger was supported in part by the National Science Foundation under Agreement No. 0112050.

1 Introduction

Complex oscillatory behaviour known as mixed mode oscillations (MMO's) were first discovered in the famous Belousov-Zhabotinsky reaction [29]. Since then, MMO's have been frequently observed in both experiments (see e.g. [17, 25, 14, 6, 20]) and models (see e.g. [8, 13, 7, 18, 21]) of chemical and biological systems. The oscillatory behaviour exhibited by these nonlinear systems consists of L large amplitude oscillations followed by s small amplitude oscillations and the symbol L^s is assigned to that pattern. As a function of a control parameter, MMO's often follow bifurcation sequences known as Farey sequences [4]. This ubiquitous behaviour suggests the possibility of the existence of a common dynamical systems theory origin.

Among the proposed mechanisms for MMO's are break-up of an invariant torus [16] and break up/loss of stability of a Shilnikov homoclinic orbit [1, 13]. Both of these scenarios are consistent with the occurence of Farey sequences, but there are certain features of MMO's which cannot be explained. For example the theory of a Shilnikov homoclinic predicts a bifurcation sequence of a growing number of small excursions which has to terminate in the formation of the homoclinic connection corresponding to an infinite amount of small oscillations. However, the most typically observed bifurcation sequences of MMO's are characterized by a bounded number of small oscillations, with the MMO behavior terminating suddenly, without the formation of a homoclinic orbit.

In this paper we present another possible explanation for MMO's. This mechanism is based on a canard phenomenon which can be observed in a 3D framework of slow-fast systems [19, 26, 28]. The classical canard phenomenon (discovered by a group of French mathematicians [2]) occurs in systems with one slow and one fast variable. Its nature is the transition from a small amplitude oscillatory state to a (large amplitude) relaxation oscillatory state within an exponentially small range of the control parameter. This transition, also called *canard explosion*, occurs through a sequence of canard cycles which can be asymptotically stable, but they are very hard to observe in experiment or simulation because of sensitivity to the control parameter and also because of sensitivity to noise. This is well known in the chemical literature where canard explosion is classified as a hard transition, because, for practical purposes, the transition from a small cycle to a relaxation oscillation occurs immediately.

Very important for the understanding of canard explosion is a trajectory called *maximal canard*, which is a connection from the stable slow manifold to the unstable slow manifold. In 2D systems maximal canards occur for discrete values of the control parameter [15], but in 3D maximal canards are robust and are therefore persistent under small parameter changes [26]. It was shown in [28] that a class of canards in 3D called canards of *folded node type* can be responsible for the small amplitude oscillations observed in MMO's. A good intuition is that the system moves dynamically from the excitable state to the oscillatory state and the feature of the large oscillation is to bring the system back to the excitatory state. The number of small oscillations predicted by this canard phenomenon is finite, which is a major difference to the mentioned theory of Shilnikov homoclinic orbit. Furthermore, this canard phenomenon is robust under noise in the sense that many of the deterministically obsrved MMO patterns are preserved.

Another attractive feature of the canard mechanism for MMO's is that, once the slow and the fast variables have been identified, it is very easy to verify if it is the canard based mechanism for MMO's or not. For example, for the systems of [7], [18] and [21, 22] it is known that the MMO mechanism is of canard type.

The main purpose of this work is to explain how MMO's can occur in systems with two slow variables due to the canard phenomenon. We want to point out the main local structure creating the small amplitude oscillations as well as the conditions a global return mechanism must satisfy for MMO's to arise. Furthermore we would like to give some insight to how Farey sequences can be explained by this mechanism. This paper is partly expository and non-technical, but there are more technical parts where we try to explain certain crucial features of the mechanism in quite some detail (in fact the paper does contain some new results). We try to avoid very technical arguments, avoiding the use of the blow-up method, which has been the main technical tool in much of our work. (For an introduction to the use of the blow up method in geometric singular perturbation theory we refer the reader to [15].) Our intention is that this paper should be accessible to anyone familiar with basic dynamical systems and singular perturbation theory, but it should also be useful for people who study MMO's in their research and are not necessarily familiar with the blow-up method.

It is worth mentioning that canard solutions play a fundamental role in other very interesting contexts. Two most notable examples are the dynamics of the forced van der Pol oscillator [10, 3] and a mechanism for localization of oscillation in chemical systems [22, 23, 24] in which two oscillators (or oscillatory clusters) are in different amplitude regimes with an order of magnitude difference between them.

The outline of the paper is as follows: in Section 2 we give a non-technical overview of the mechanism of mixed mode oscillations we consider. In Section 3 we present some results on the dynamics near folded nodes, concentrating on the information relevant for MMO's. This section contains the main new result of this paper, namely the estimates of the size of the sectors of rotation. In Section 4 we state some results on the existence of periodic orbits of type 1^s (one large and s small oscillations) and give very rough sketches of the proofs. We also indicate how periodic orbits with other Farey sequences can come about. In Section 5 we present some simulations corroborating the results of Section 4. In Section 6 we indicate how the analysis can be extended to systems with two slow and arbitrarily many fast dimensions. Finally, in Section 7 we present a method for finding mixed mode oscillations in systems with two slow and arbitrarily many fast variables.

2 Mixed mode oscillations in three dimensions

In this section we present expository material on MMO's in three dimensions. More precisely, we consider systems of the form

$$\begin{aligned} \varepsilon\dot{x} &= f(x,y,z) \\ \dot{y} &= g_1(x,y,z) \\ \dot{z} &= g_2(x,y,z)\,, \end{aligned} \tag{2.1}$$

with $x \in \mathbb{R}$ as the fast variable and $(y, z) \in \mathbb{R}^2$ as slow variables. We also consider the fast system, which is equivalent to system (2.1) on the fast time scale $\tau = t/\varepsilon$,

$$\begin{aligned} x' &= f(x, y, z) \\ y' &= \varepsilon g_1(x, y, z) \\ z' &= \varepsilon g_2(x, y, z) \,. \end{aligned} \tag{2.2}$$

The two limits obtained by setting $\varepsilon = 0$ in (2.1) and (2.2), respectively, are known as the *reduced problem* and the *layer problem.* The basic philosophy of geometric singular perturbation theory is to characterize solutions of the full problem using the information on both limiting subsystems. Let $S = \{(x, y, z) \,:\, f(x, y, z) = 0\}$ denote the critical manifold which is the phase space of the reduced problem.

Assumption 1 *The critical manifold S is (locally) a folded surface, i.e. $S = S_a \cup \mathcal{F} \cup S_r$ with attracting branch $S_a := \{(x, y, z) \in S \,:\, f_x < 0\}$, repelling branch $S_r := \{(x, y, z) \in S \,:\, f_x > 0\}$ and fold-line $\mathcal{F} := \{(x, y, z) \in S \,:\, f_x = 0,\, f_{xx} \neq 0\}$*

Without loss of generality, we assume that $f_y|_{p \in \mathcal{F}} \neq 0$ holds locally. It follows by the implicit function theorem that S is locally given as a graph $y = \phi(x, z)$, i.e. $f(x, \phi(x, z), z) = 0$. We use this representation $y = \phi(x, z)$ to define a projection of the reduced system

$$\begin{aligned} 0 &= f(x, y, z) \\ \dot{y} &= g_1(x, y, z) \\ \dot{z} &= g_2(x, y, z) \end{aligned} \tag{2.3}$$

onto S. Implicit differentiation of $f(x, y, z) = 0$ gives the relationship $-f_x \dot{x} = (f_y g_1 + f_z g_2)$. It follows that the reduced problem (2.3) projected onto the (x, z)-plane is given by

$$\begin{pmatrix} -f_x \, \dot{x} \\ \dot{z} \end{pmatrix} = \begin{pmatrix} (f_y \, g_1 + f_z \, g_2) \\ g_2 \end{pmatrix} \Bigg|_{y = \phi(x, z)} . \tag{2.4}$$

Note that the fold-line $\mathcal{F}$ is the locus of singularities of (2.4). We rescale time by the factor $-f_x$, obtaining the desingularized system of (2.4):

$$\begin{pmatrix} \dot{x} \\ \dot{z} \end{pmatrix} = \begin{pmatrix} (f_y \, g_1 + f_z \, g_2) \\ -f_x \, g_2 \end{pmatrix} \Bigg|_{y = \phi(x, z)} . \tag{2.5}$$

The local dynamics near $\mathcal{F}$ of system (2.5) can be completely understood. The phase portraits of the reduced problem (2.4) and the desingularized system (2.5) are the same, but the orientation of the flow has to be reversed on S_r, see Figure 1.

Typically, fold points $p \in \mathcal{F}$ satisfy the normal switching condition

$$(f_y \, g_1 + f_z \, g_2)|_{p \in \mathcal{F}} \neq 0. \tag{2.6}$$

A fold point p satisfying (2.6) is a *jump point* if there exists a trajectory of (2.4) starting at some point q in S_a and ending at p. Jump points play a fundamental role in relaxation oscillations [27].

Under condition (2.6) the reduced system (2.4) becomes unbounded along the fold-line $\mathcal{F}$. Thus trajectories of system (2.1) reaching the vicinity of $\mathcal{F}$ subsequently jump away from the fold, i.e. a fast transition away from the critical manifold S near $\mathcal{F}$ almost parallel to the x-axis occurs. This jumping behaviour near $\mathcal{F}$ is part of the mechanism leading to relaxation oscillations in system (2.1).

Note that away from $\mathcal{F}$ there exist, by Fenichel theory [9, 12], an attracting manifold $S_{a,\varepsilon}$ and a repelling manifold $S_{r,\varepsilon}$, with S_a and S_r as their singular limits. $S_{a,\varepsilon}$ is only locally invariant and trajectories arrive in the vicinity of the fold-line $\mathcal{F}$ after finite time, where the slow flow on $S_{a,\varepsilon}$ is approximately described by the reduced flow (2.4). Then solutions are forced to jump away from the fold. If, after this fast transition away from the fold, a global return mechanism projects the trajectory back onto the attracting manifold $S_{a,\varepsilon}$, then periodic solutions are possible and we obtain relaxation oscillations. This oscillations can be viewed as simple 1^0 patterns of MMO's. For details on relaxation oscillations we refer to [27, 11].

To obtain true MMO's including small amplitude oscillations the reduced flow has to possess a folded singularity. Typically, a folded singularity is an isolated point $p_0 \in \mathcal{F}$ which violates the normal switching condition (2.6). Therefore, a folded singularity is an equilibrium of the desingularized system (2.5). We call p_0 a *folded node, folded saddle, or folded saddle-node* if, as an equilibrium of (2.5), it is a node, a saddle, or a saddle-node.

Remark: There are two very important limits of folded node, called folded saddle-node of type I and folded saddle-node of type II, both of codimension one in the class of systems we consider. A *folded saddle node* of type I occurs when the linearization of (2.5) at p_0 has an eigendirection tangent to $\mathcal{F}$. Note that $\mathcal{F}$ is a nullcline of (2.5), which implies that the corresponding eigenvalue must be 0.[1] The occurence of the 0 eigenvalue explains the use of the term 'saddle-node'. In fact a generic folded saddle-node of type I leads to the birth of a folded node/folded saddle pair.

A *folded saddle node* of type II corresponds to the case when the linearization of (2.5) at p_0 has a zero eigenvalue with the eigendirection transverse to the fold line $\mathcal{F}$. In the unfolding of a folded saddle node of type II an equilibrium passes from the stable to the unstable sheet of the slow manifold. See [26, 28] for more details on folded saddle nodes.

Assumption 2 *The reduced system (2.3) possesses a folded node singularity $p_0 \in \mathcal{F}$, which is defined by the condition*

$$(f_y g_1 + f_z g_2)|_{p_0 \in \mathcal{F}} = 0 \tag{2.7}$$

and by the requirement that the eigenvalues λ_w and λ_s of the linearization of (2.5) at p_0 satisfy $\lambda_w \cdot \lambda_s > 0$. In addition we assume that λ_w and λ_s are negative.

Henceforth we assume that Assumptions 1 and 2 hold for (2.1) respectively (2.2).

Canards are special trajectories which play a very important role in the dynamics near a folded node. Their defining characteristic is that they originate in $S_{a,\varepsilon}$, continue to the vicinity of the fold, and subsequently, instead of jumping along the fast direction, continue to $S_{r,\varepsilon}$. In geometric terms, a canard corresponds to the intersection of the manifolds $S_{a,\varepsilon}$ and $S_{r,\varepsilon}$ extended by the flow to the vicinity of the folded singularity. Canards are not unique because the corresponding invariant manifolds $S_{a,\varepsilon}$ and $S_{r,\varepsilon}$ are not unique (but exponentially close). For a fixed choice of the invariant manifolds we call their intersections *maximal canards*. Each

[1] In general, if a system of differential equations has an equilibrium with an eigendirection tangent to one of the nullclines then the corresponding eigenvalue must be 0.

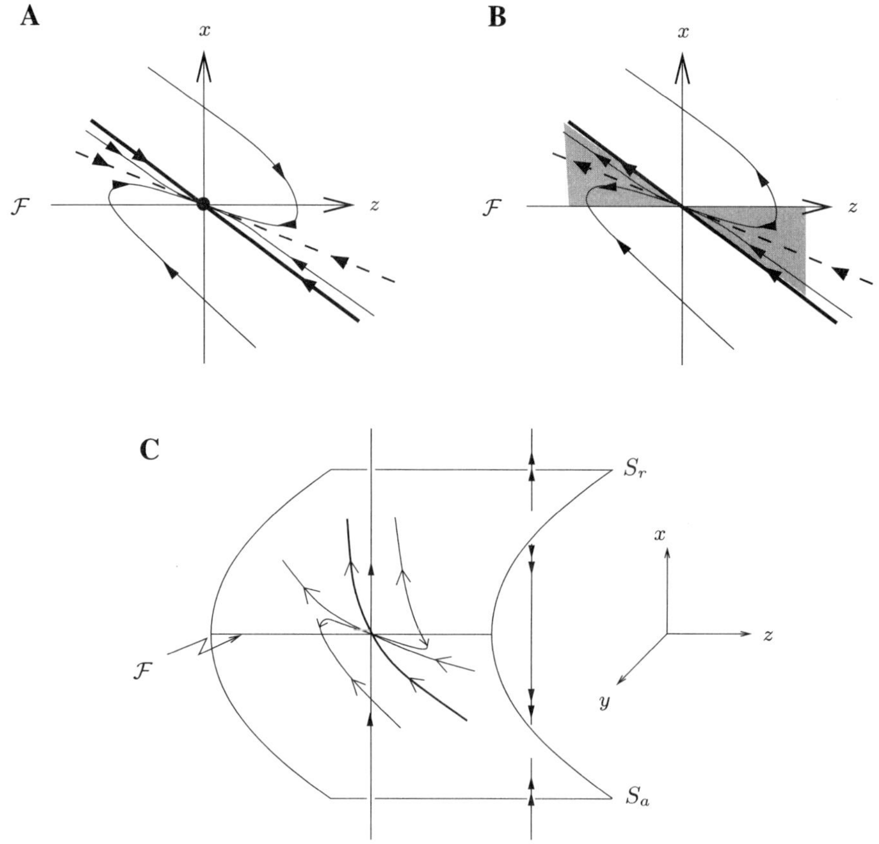

Figure 1 Folded node, A: the desingularized flow, the bold resp. dashed line corresponds to the strong resp. weak eigendirection of the singularity, B: the reduced flow with the singular funnel (shadowed region), C: the reduced flow on the critical manifold S

maximal canard defines a family of canards nearby which are exponentially close to the maximal canard. *Singular canards*, or *canard candidates* are trajectories of the reduced flow (2.4) which pass at p_0 from S_a to S_r. For a folded node there exists a whole sector of singular canards, see Figure 1. Canards existing for (2.1) for $\varepsilon > 0$ arise as perturbations of singular canards. Their number depends in an intricate way on the parameter $\mu = \lambda_w/\lambda_s$ [28]. (we assume WLOG that $|\lambda_w| < |\lambda_s|$). Two of the maximal canards, called the primary canards, are perturbations of the eigendirections of λ_w and λ_s of the desingularized system (2.5). The primary canard corresponding to the eigendirection of λ_w is called the *weak canard* and the primary canard corresponding to the eigendirection of λ_s is called the *strong canard*. It was shown in [28] that near a folded node singularity the trajectories of system (2.1) originating close to the singular funnel enter a region near the fold where they rotate about the weak canard. The rotational property and the existence of the weak canard can be explained purely by a geometric argument. The strong canard can be thought of the locus of a change in the relative position of $S_{a,\varepsilon}$ and $S_{r,\varepsilon}$. The trajectories originating in $S_{a,\varepsilon}$ to one side of the strong canard are trapped by $S_{r,\varepsilon}$ and have to return towards $S_{a,\varepsilon}$. This causes the small rotations and the weak canard can be understood as the center of rotation. The trajectories originating in $S_{a,\varepsilon}$ to the other side of the strong canard are prevented by $S_{r,\varepsilon}$ from returning to $S_{a,\varepsilon}$ and must follow a fast direction. We refer to the region of rotation as the *funnel*. During their passage near the funnel the trajectories are initially attracted to the weak canard and later repelled from it. The trajectories which enter the funnel are trapped for a significant amount of time, however this behavior is transient and eventually the trajectory exits the vicinity of the fold by jumping along the fast direction. If, after the jump away, a global return mechanism projects this trajectory back onto $S_{a,\varepsilon}$ within the area of attraction of the funnel, then MMO patterns of type 1^s are possible, i.e. periodic solutions consisting of one large amplitude oscillation due to the global return mechanism and s small amplitude oscillations due to the funnel near the folded node singularity.

We would like to describe MMO's in their singular limit. Note that for small $\varepsilon > 0$ fast jumps are executed near the folded node singularity. In the singular limit we describe all these jumps as a projection along the fast fiber of the layer problem through the folded node onto a possible other attracting critical manifold where the trajectory then follows the local reduced flow until it may reach another fold where it gets projected back along another fast fiber onto the original attracting manifold. Cubic shaped critical manifolds are a prominent example which lead to this return mechanism and they are ubiquitous in chemical and biological applications. With that description of the fast jumps we can define a singular limit of MMO's:

Assumption 3 *There exists a singular periodic orbit $\Gamma = \Gamma_a \cup \Gamma_g$ for system (2.1) which consists of a smooth segment Γ_a on S_a within the singular funnel with the folded node singularity p_0 as endpoint. Γ_g represents a global return mechanism onto S_a which has to be specified.*

Theorem 2.1 *Given system (2.1) under Assumption 1-3.*

1. *There exist mixed mode oscillations of type 1^s. It is possible to calculate the maximal number of small oscillations s.*
2. *If the segment Γ_a of the singular periodic orbit Γ consists of a segment of the boundary of the singular funnel (the strong canard) then more complicated MMO patterns of type $L^{s'}$, $L \geq 1$ and $s' \leq s$, are possible.*

Figure 2 shows the basic mechanism to obtain MMO's. In the following we discuss the local and the global aspects to obtain MMO patterns in more detail. We define a return map $\Pi : \Sigma^- \to \Sigma^-$, where Σ^- is a cross section orthogonal to the x-axis away from the fold, such that for small ε, all trajectories under consideration which are projected back onto $S_{a,\varepsilon}$ onto the neighbourhood of the funnel intersect Σ^- (see Figure 3). We decompose this map into $\Pi := \Pi_G \circ \Pi_L$, where $\Pi_L : \Sigma^- \to \Sigma^+$ defines the (local) map near the folded critical manifold, and $\Pi_G : \Sigma^+ \to \Sigma^-$ defines the (global) map defined by the global return mechanism. We begin our analysis with the map Π_L.

3 Small amplitude oscillations – the folded node

The map $\Pi_L : \Sigma^- \to \Sigma^+$ captures the dynamics near the folded singularity where canard solutions exist. Recall that canard solutions are trajectories originating on $S_{a,\varepsilon}$, passing through the neighborhood of a folded singularity, and ending up in $S_{r,\varepsilon}$ where they remain for a considerable amount of time. In [28] the existence of *primary and secondary canards* (see Figure 6) due to a folded node singularity

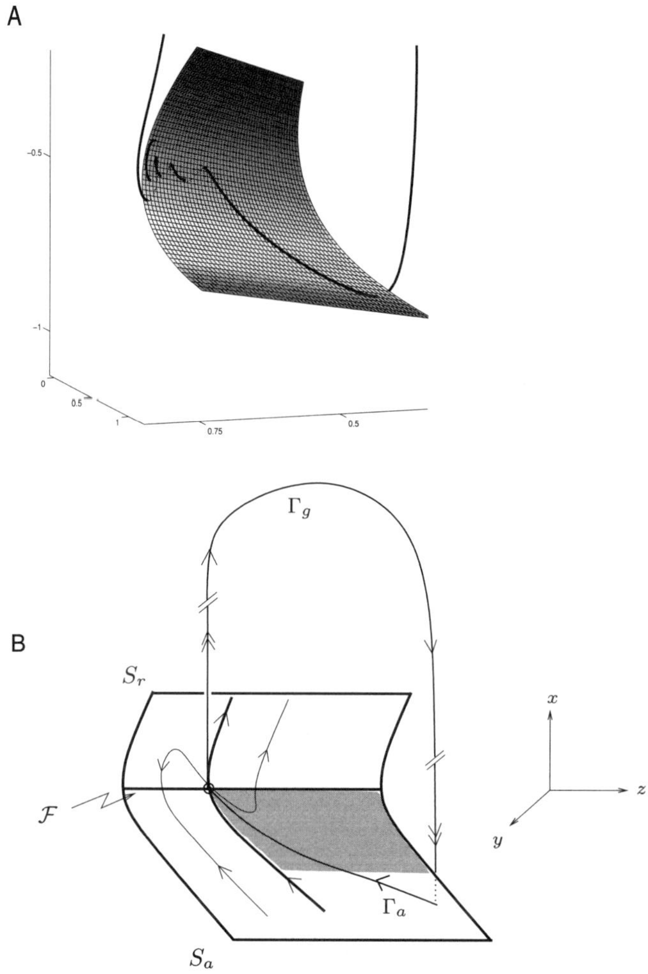

Figure 2 The basic mechanism to generate MMO patterns: A folded node singularity (B: black circle) forms a singular funnel (B: shadowed region). A trajectory Γ_a which enter this funnel makes small amplitude oscillations near the folded node singularity. After leaving the funnel the trajectory may jump away along a fast direction. A global return mechanism (Γ_g) injects the trajectory back into the funnel forming a periodic solution with a 1^s MMO pattern.

was shown as well as rotational properties of solutions near the folded singularity. In the following we analyse these small amplitude oscillations in more detail.

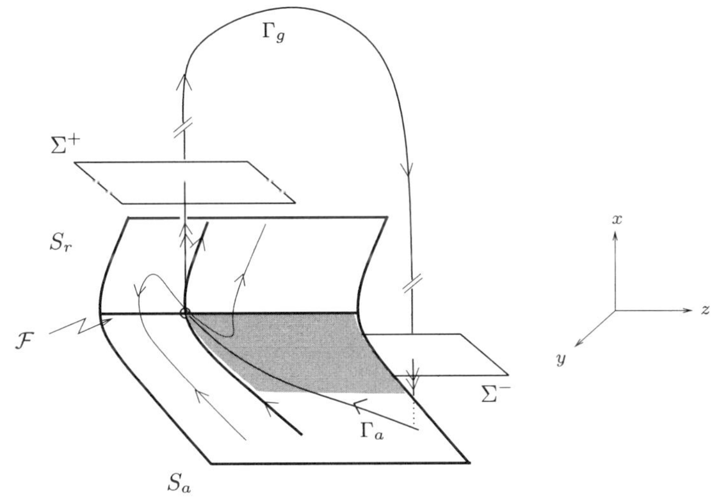

Figure 3 Cross sections Σ^- and Σ^+ to analyse MMO patterns.

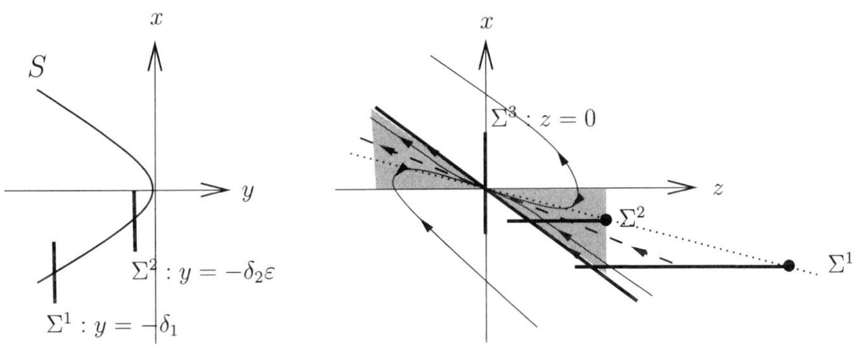

Figure 4 Left: cross-sections Σ^1 and Σ^2 near the fold; Right: projection of reduced flow onto (x, z) plane. Note that the projection of the intersection of the cross-section Σ^1 with the invariant manifold S is one-dimensional. Σ^1 intersects the line of turning points (dotted line). Therefore Σ^1 is just a cross section of the reduced flow to the left of the intersection (black dot) where the flow is directed towards the fold. The same argument holds for the definition of the cross section Σ^2.

Remark: Note that due to the existence of canards the map $\Pi_L : \Sigma^- \to \Sigma^+$ may not be well defined. In the following analysis it will become clear which restrictions have to be made to define the map Π_L well.

Fenichel theory in its classical form guarantees the existence of the slow manifolds $S_{a,\varepsilon}$ and $S_{r,\varepsilon}$ $O(1)$ away from the fold. Thus we define a map $\Pi_1 : \Sigma^- \to \Sigma^1$ where Σ^1 is a cross section defined by $y = -\delta_1$, $\delta_1 > 0$ small, parallel to the fast fibers and $O(1)$ away from the fold. Note there is a natural restriction for Σ_1 to be a cross section due to the folded node. As shown in Figure 4, the flow in a neighbourhood of a folded node has turning points (shown as a dotted line) with respect to z , i.e. the vector field at turning points is parallel to the z-axis. Therefore

Σ^1 may be just well defined as a cross section on a certain interval I_z of z. But this interval I_z always covers the area between the two eigendirections of the folded node (shown as bold and dashed lines) which becomes important in the following analysis.

We next turn to the flow of trajectories past the folded node. We introduce normalizing coordinates which brings system (2.1) into a canonical form near a folded node singularity given by

$$\begin{aligned} \varepsilon\dot{x} &= y + x^2 + O(x^3, x^2 z, xyz, \varepsilon) \\ \dot{y} &= -(\mu+1)x - z + O(y, \varepsilon, (x+z)^2) \\ \dot{z} &= -\frac{\mu}{2}, \qquad \mu \in (0,1). \end{aligned} \tag{3.1}$$

Remark: System (3.1) can be seen as one possible normal form near a folded node. The limit $\mu \to 0$ gives the folded saddle-node type II. In [28] a similar normal form of the folded node was studied which has a folded saddle-node type I as its limit for $\mu \to 0$. The results on folded nodes are the same for both normal forms, just the limiting behavior of the folded saddle-nodes is different. Therefore all results on folded nodes obtained in [28] are valid for system (3.1).

The critical manifold S of system (3.1) is a parabolic cylinder with the fold-line $\mathcal{F}$ given by the z-axis and the folded singularity p_0 is at the origin. The reduced system for (3.1) is

$$\begin{aligned} -2x(1+O(x,z))\dot{x} &= -(\mu+1)x - z + O((x+z)^2) \\ \dot{z} &= -\mu/2 \end{aligned} \tag{3.2}$$

and the desingularized system is

$$\begin{aligned} \dot{x} &= -(\mu+1)x - z + O((x+z)^2) \\ \dot{z} &= \mu x + O(x^2, xz)\,. \end{aligned} \tag{3.3}$$

The origin is an equilibrium of node type with eigenvalues $\lambda_s = -1$ and $\lambda_w = -\mu$. Thus $\mu = \lambda_w/\lambda_s \in (0,1)$ defines the ratio between the strong and the weak eigenvalue of the folded node. The phase portraits of the reduced system (3.2) and the desingularized system (3.3) are equivalent to those shown in Figure 1.

We are studying the map $\Pi_f : \Sigma^1 \to \Sigma^+$ induced by the flow of system (3.1). To analyse this map past the folded node we have to introduce intermediate (singular) cross-sections which are sections used in the blow-up analysis of the folded node in [28]. We want to avoid technicalities in this paper and therefore will not introduce the blow-up analysis, but refer for details to [28].

To understand the flow past the fold-line we have to rescale system (3.1) by setting

$$x = \sqrt{\varepsilon}\bar{x}, \quad y = \varepsilon\bar{y}, \quad z = \sqrt{\varepsilon}\bar{z}, \quad t = \sqrt{\varepsilon}\bar{t}. \tag{3.4}$$

Equations (3.1) transform to

$$\begin{aligned} \bar{x}' &= \bar{y} + \bar{x}^2 + O(\sqrt{\varepsilon}) \\ \bar{y}' &= -(\mu+1)\bar{x} - \bar{z} + O(\sqrt{\varepsilon}) \\ \bar{z}' &= -\frac{\mu}{2}, \end{aligned} \tag{3.5}$$

where $'$ denotes $d/d\bar{t}$. System (3.5) is a zoom of the vector field near the folded node singularity. Note that the section Σ^1 is given by $\bar{y} = -\frac{\delta_1}{\varepsilon}$ in system (3.5). We introduce a section $\Sigma^2 : \bar{y} = -\delta_2$, with sufficiently large $\delta_2 > 0$, and define the map $\Pi_2 : \Sigma^1 \to \Sigma^2$.

Remark: Σ^2 is given by $y = -\delta_2\varepsilon$ in the original system (3.1). Similar to Σ^1, the section Σ^2 is just a cross section for a certain interval $I_{z'}$ of z which covers the area of the strong and the weak eigendirection of the folded node. Under the reduced flow (3.2) the interval I_z of Σ^1 is mapped into the interval $I_{z'}$ of Σ^2 (see Figure 4). Therefore the map Π_2 is well defined.

For $\bar{y} < -\delta_2$, it was shown in [28] that system (3.5) has still a normally hyperbolic structure and Fenichel theory can be applied. More specifically, the manifold $S_{a,\varepsilon}$ given by classical Fenichel theory in section Σ^1 can be extended up to the section Σ^2 by the flow of (3.5). Since the extended manifold is an invariant manifold of (3.5) we denote it by $S_{a,\sqrt{\varepsilon}}$. A manifold $S_{r,\sqrt{\varepsilon}}$ is obtained analogously, except by going backwards in time. The following result holds:

Proposition 3.1 [28] *For system (3.5) with $\varepsilon \neq 0$ and also for (3.1) the sets $S_{a,\sqrt{\varepsilon}}$ and $S_{r,\sqrt{\varepsilon}}$ are smooth locally invariant normally hyperbolic manifolds and $O(\sqrt{\varepsilon})$ smooth perturbations of S. The flow on $S_{a,\sqrt{\varepsilon}}$ (respectively on $S_{r,\sqrt{\varepsilon}}$) is an $O(\sqrt{\varepsilon})$ perturbation of the reduced flow.*

Remark: Proposition 3.1 is an extension Fenichel theory up to $O(\sqrt{\varepsilon})$ from the fold $\mathcal{F}$ (which corresponds to $O(\varepsilon)$ in the y direction). In fact it can be proved by application of invariant manifold theory in suitable coordinates [26]. Therefore, the map Π_2 can be completely described by Fenichel theory.

Henceforth we focus on $S_{a,\sqrt{\varepsilon}}$ and $S_{r,\sqrt{\varepsilon}}$ in the coordinates $(\bar{x}, \bar{y}, \bar{z})$ of system (3.5). The advantage of the 'blown-up' system (3.5) is that we can take now the limit $\varepsilon \to 0$ and obtain sets S_a and S_r which are locally invariant for $\{(3.5), \varepsilon = 0\}$. The sets S_a and S_r can be seen as the approximations of $S_{a,\sqrt{\varepsilon}}$ and $S_{r,\sqrt{\varepsilon}}$ at the $O(\sqrt{\varepsilon})$ level. In the following we want to understand how S_a and S_r behave beyond Σ^2. One important property is that S_a, respectively S_r, consist of solutions of $\{(3.5), \varepsilon = 0\}$ of at most algebraic growth as $\bar{t} \to -\infty$, respectively $\bar{t} \to \infty$. In fact these manifolds approach the parabolic cylinder $\bar{y} = -\bar{x}^2$ for $\bar{t} \to \pm\infty$. Furthermore, the unperturbed system $\{(3.5), \varepsilon = 0\}$ has two explicitly known polynomial solutions $\varphi_w(\bar{t})$ and $\varphi_s(\bar{t})$:

$$\varphi_w(\bar{t}) = \left(\frac{\mu}{2}\bar{t},\ -\frac{\mu^2}{4}\bar{t}^2 + \frac{\mu}{2},\ -\frac{\mu}{2}\bar{t}\right)$$

$$\varphi_s(\bar{t}) = \left(\frac{1}{2}\bar{t},\ -\frac{1}{4}\bar{t}^2 + \frac{1}{2},\ -\frac{\mu}{2}\bar{t}\right)$$

We refer to $\varphi_w(\bar{t})$ and $\varphi_s(\bar{t})$ as *singular canards* in the context of the rescaled variables (3.4) and the corresponding system $\{(3.5), \varepsilon = 0\}$. Recall that we defined *singular canards* for (2.4) as solutions that continue from S_a to S_r. Furthermore, $\varphi_w(\bar{t})$ and $\varphi_s(\bar{t})$ are related to the strong and weak eigendirection of the folded node (these are obtained by projecting $\varphi_w(\bar{t})$ and $\varphi_s(\bar{t})$ onto the (x, z) plane). The solutions $\varphi_w(\bar{t})$ and $\varphi_s(\bar{t})$ play a very important role in understanding the properties of $S_{a,\sqrt{\varepsilon}}$ and $S_{r,\sqrt{\varepsilon}}$ beyond Σ^2.

Proposition 3.2 [26] *Consider (3.5) with* $\varepsilon = 0$. *The invariant sets* S_a *and* S_r *are smooth 2 dimensional submanifolds with a common boundary, given by* $\varphi_w(\bar{t})$. *They have first order contact along* $\varphi_w(\bar{t})$ *if* $(1/\mu) \notin \mathbb{N}$ *and higher order contact at* $\varphi_w(\bar{t})$ *if* $(1/\mu) \in \mathbb{N}$. *In addition there is at least one intersection, given by* $\varphi_s(\bar{t})$. *This intersection is transverse for every* $\mu \in (0,1)$.

Remark: It is clear that $\varphi_w(\bar{t})$ and $\varphi_s(\bar{t})$ are both contained in S_a and in S_r (see Figure 4). However the fact that $\varphi_w(\bar{t})$ is the common boundary of S_a and S_r is not immediately clear and requires an additional estimate. Later in this Section we will develop some technical tools which will enable us to sketch the proof of this result.

By a standard application of the Implicit Function Theorem each transverse intersection of S_a and S_r perturbs to a transverse intersection of $S_{a,\sqrt{\varepsilon}}$ and $S_{r,\sqrt{\varepsilon}}$, which corresponds to a maximal canard solution. Thus Proposition 3.2 establishes the existence of the strong canard for any $\mu \in (0,1)$ and the existence of the weak canard for non-resonant μ.

We now consider additional intersections of S_a and S_r. Note that the manifolds S_a and S_r rotate around the singular weak canard. These rotations happen on a compact domain. Rotational properties of the tangent bundles along the weak canard ϕ_w can be inferred from the properties of the solutions of the Weber equation $v'' - tv' + v/\mu = 0$. More specifically, the following holds.

Proposition 3.3 [28] *Given system (3.5) with* $\varepsilon = 0$ *and* $n-1 < 1/\mu < n$, *the invariant manifolds* S_a *and* S_r *twist* $(n-1/2)$ *times around the primary weak canard in the neighbourhood of the folded node singularity, where a twist corresponds to a half rotation of* $180°$.

Let Σ^3 be a cross section of the flow of (3.5) defined by $\bar{z} = 0$. Note that intersection points of the curves $S_a \cap \Sigma^3$ and $S_r \cap \Sigma^3$ are in one-to-one correspondance with solutions along which S_a and S_r intersect. In the resonant cases $1/\mu \in \mathbb{N}$, the curves $S_a \cap \Sigma^3$ and $S_r \cap \Sigma^3$ have higher order contact at $\phi_w(0)$. Odd resonances lead to a new intersection point while passage through even resonances leads to no new intersections. Wechselberger [28] proves the following result.

Proposition 3.4 [28] *Consider system (3.5) with* $\varepsilon = 0$. *Suppose that* $k > 0$ *is an integer such that*

$$2k + 1 < \frac{1}{\mu} < 2k + 3.$$

Suppose in addition that $(1/\mu) \neq 2(k+1)$. *Then* $S_a \cap \Sigma^3$ *and* $S_r \cap \Sigma^3$ *have* k *transverse intersection points in addition to* $p_s(0)$ *and the contact point* $p_w(0)$.

Figure 5 shows 2+2 intersections of $S_a \cap \Sigma^3$ and $S_r \cap \Sigma^3$ of system $\{(3.5), \varepsilon = 0\}$ for $\mu = 0.15$.

Remark: Due to the time reversing symmetry of the unperturbed system $\{(3.5), \varepsilon = 0\}$ given by $T(\bar{x}, \bar{y}, \bar{z}, \bar{t}) = (-\bar{x}, \bar{y}, -\bar{z}, -\bar{t})$ the curves $S_a \cap \Sigma^3$ and $S_r \cap \Sigma^3$ are symmetric images by a reflection across the y axis.

From Proposition 3.4 we conclude

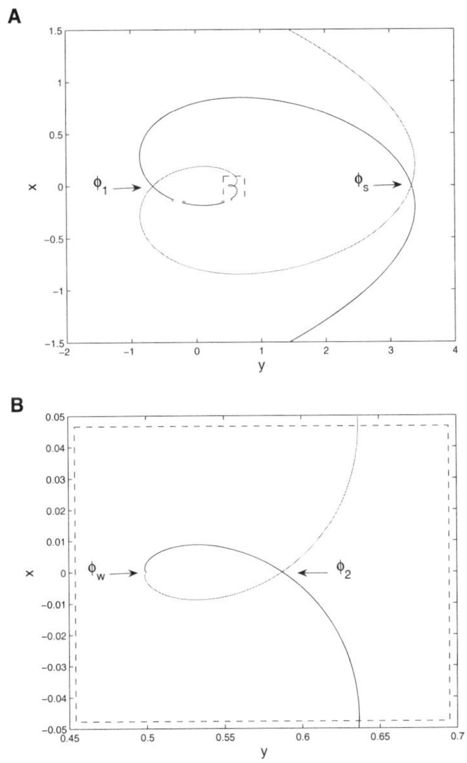

Figure 5 A: The relative position of the manifolds S_a (thin grey curve) and S_r (bold black curve) are shown in section $\Sigma^3 : \bar{z} = 0$ for $\mu = 0.15$, B: zoom of the dashed box area of A. The intersections represent the 2 primary canards denoted as ϕ_w and ϕ_s, as well as 2 secondary canards denoted by ϕ_1 and ϕ_2 predicted by Corollary 3.1. Note, the manifolds rotate around the primary weak canard ϕ_w.

Corollary 3.1 *Under the assumptions of Proposition 3.4 the system (3.1) has, for sufficiently small* ε, *k secondary maximal canards.*

Remark: The bifurcations of canards for $\varepsilon \neq 0$ are much more complicated than the bifurcations of the unperturbed system, as described in the paragraph following Proposition 3.3 (see [28]). In this paper we assume that μ is not close to resonance, where such bifurcations are of relevance.

We now describe the rotational structure of the secondary canards. For $1/\mu \geq 2k+1$, $k \in \mathbb{N}$ there exist k secondary canards positioned between the two primary canards in section Σ_2. Furthermore, the j-th secondary canard makes $(2j+1)$ twists around the primary weak canard, $1 \leq j \leq k$. The primary strong canard makes also one twist around the primary weak canard. Thus the primary strong canard and the k secondary canards are consecutively separated by a full rotation of $360°$. This allows us to define *sectors* in $S_{a,\sqrt{\varepsilon}}$ with distinctive rotational properties bordered by the primary and the secondary canards.

Let $p_j \in S_{a,\sqrt{\varepsilon}}$, $1 \leq j \leq k$, denote the intersection of the j-th secondary canard with Σ^2 and $p_w \in S_{a,\sqrt{\varepsilon}}$, resp. $p_s \in S_{a,\sqrt{\varepsilon}}$, the intersection of the primary

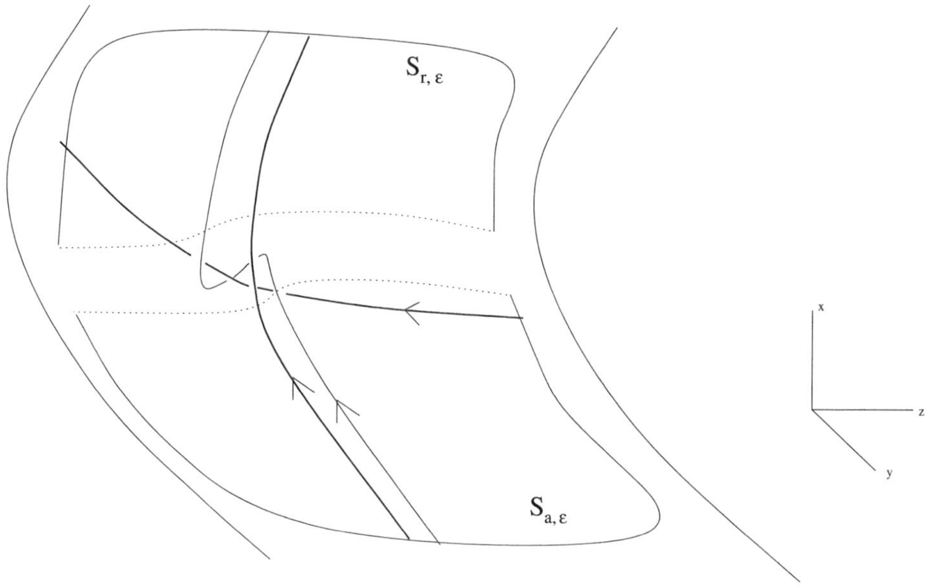

Figure 6 The two primary canards and the first secondary canard

weak canard, resp. strong canard, with Σ^2. Let I_{k+1} be the interval in $S_{a,\sqrt{\varepsilon}} \cap \Sigma^2$ bounded by the end point of $J_{z'}$ close to the weak canard and by p_k. Let I_k be the interval in $S_{a,\sqrt{\varepsilon}} \cap \Sigma^2$ bounded by p_k and p_{k-1}, I_{k-1} the interval in $S_{a,\sqrt{\varepsilon}} \cap \Sigma^2$ bounded by p_{k-1} and p_{k-2}, and so on. Finally, I_1 is the interval bounded by p_1 and p_s. Trajectories with initial conditions in the interior of I_j, $1 \leq j < k+1$, make $(2j + 1/2)$ twists around the primary weak canard, while trajectories with initial conditions in the interior of I_{k+1} make at least $[2(k+1) - 1/2]$ twists around the primary weak canard. All these solutions are forced to follow the *funnel* created by the manifolds $S_{a,\sqrt{\varepsilon}}$ and $S_{r,\sqrt{\varepsilon}}$. After solutions leave the funnel they get repelled by the manifold $S_{r,\sqrt{\varepsilon}}$ and will follow close to a fast fiber in the $\bar{x}$ direction and exit finally through the section Σ^+ given by $\bar{x} = \delta_+/\sqrt{\varepsilon}$ in system (3.5). The singular limit of this fast fiber for the original problem (3.1) is exactly the fiber through the folded node singularity. Note that all these trajectories do not exit through Σ^2.

Remark: The boundaries of the sectors are primary and secondary maximal canards. Consequently, the trajectories close to the boundaries of the sectors will exit consecutively through the sections Σ^2 and Σ^1 near S_r and follow the slow flow near S_r close to the corresponding maximal canard for a certain amount of time before they get repelled and jump away (see Figure 6). Depending on the specific problem under study the position of the exit section Σ^+ may not be suitable, or has to be modified to capture the trajectories passing very close to the boundaries of the sectors.

The remaining part of this section is devoted to estimating the size of the sectors of rotation. More specifically, we continue the secondary canards backwards in time on $S_{a,\varepsilon}$ all the way to the section Σ_1 and ask for the interval size I'_j defined by the canards in Σ_1. The size of the intervals I_j, $1 \leq j \leq k+1$, in the section Σ^2 of system

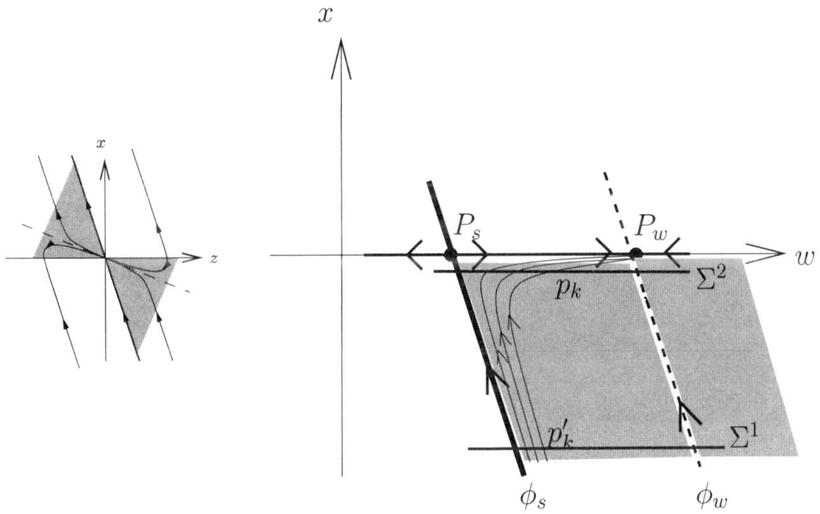

Figure 7 'Blown-up' reduced flow obtained by introducing a new coordinate w=-z/x

(3.5) is $O(1)$, i.e. the primary and secondary canards are all finitely separated. Note that an interval size $O(1)$ in $\Sigma_2 : \bar{y} = -\delta_2$ of system (3.5) becomes an interval size $O(\sqrt{\varepsilon})$ in $\Sigma_2 : y = -\delta_2\varepsilon$ of the original system (3.1).

To estimate the size of the intervals I_j' in Σ^1 we are studying the (inverse) transition map Π_2^{-1} for the reduced flow of (3.2). Since transition maps are independent of time parametrization, we can use the desingularized system (3.3) instead of the reduced system (3.2). In system (3.3) the section Σ^1 is approximately given by $x = -\sqrt{\delta_1}$ and the section Σ^2 is approximately given by $x = -\sqrt{\delta_2\varepsilon}$. To understand the reduced flow close to the node singularity, we introduce a new variable $w = -z/x$. This singular coordinate change unfolds the node singularity and resembles the blow-up analysis done in [26, 28]. In the new coordinates (x, w) the desingularized system (3.3) becomes

$$\begin{aligned} \dot{x} &= -x(\mu + 1 - w + O(x)) \\ \dot{w} &= -(w - \mu)(w - 1) + O(x). \end{aligned} \tag{3.6}$$

Note that the sections $\Sigma^1 : x = -\sqrt{\delta_1}$ and $\Sigma^2 : x = -\sqrt{\delta_2\varepsilon}$ are still the same, but the size of intervals I_j change to $O(1)$ in Σ^2 for system (3.6). Thus, the 'blown-up' desingularized system (3.6) approximates the slow dynamics of the rescaled system (3.5). Looking at the dynamics of (3.6) we find two equilibria: a sink $P_w = (0, 1)$ and a saddle $P_s = (0, \mu)$. The unique attracting eigendirection of the saddle P_s corresponds to the primary strong canard ϕ_s while the incoming attracting eigendirection of the sink P_w corresponds to the the primary weak canard ϕ_w. Thus the 'blow-up' splits the strong and weak eigendirection of the folded node singularity (see Figure 7). We estimate the (inverse) transition time from Σ^2 to Σ^1 and the size of the intervals I_j' in Σ^1 by the linearized system at the equilibrium P_s. The linerization at P_s is given by:

$$\begin{pmatrix} \dot{x} \\ \dot{w} \end{pmatrix} = \begin{pmatrix} -1 & 0 \\ * & (1-\mu) \end{pmatrix} \begin{pmatrix} x \\ w \end{pmatrix}, \tag{3.7}$$

where $*$ is an unspecified term. The estimate of the (inverse) transition time from Σ^2 to Σ^1 for trajectories not close to Φ_w is

$$T = -\ln\sqrt{\varepsilon} + O(1)\,.$$

Next we want to estimate the size of the intervals I'_j, $1 \leq j \leq k+1$, in Σ_1. It is clear by the phase plane of system (3.6) that all solutions within the sector bordered by the primary canards ϕ_w and ϕ_s are moving towards the strong canard ϕ_s in backward time, because the saddle P_s has the only attracting eigendirection in backward time (see Figure 7). We have an eigenvalue $(\mu-1)$ corresponding to the attracting eigendirection in backward time along the w-axis of the saddle P_s. We make a linear coordinate change which brings (3.7) for P_s into diagonal form. Then, by solving the equation $\dot{w} = (\mu-1)w$ for $t \in (0, -\ln\sqrt{\varepsilon})$, we can estimate the positions of the points p'_j in Σ^1 corrsponding to the secondary canards ϕ_j. We know that all initial conditions p_j, $1 \leq j \leq k$, are O(1) away from the equilibrium P_s corresponding to the strong primary canard ϕ_s. We estimate that all points p'_j will be approximately $O(\varepsilon^{(1-\mu)/2})$ close to the strong primary canard ϕ_s in section Σ^1 of system (3.6). By transforming back to the original variables (x,z) the estimate $O(\varepsilon^{(1-\mu)/2})$ for the size of the intervals I'_j holds also for the desingularized system (3.3). We conclude that the following result holds.

Proposition 3.5 *All secondary canards of system (3.1) are $O(\varepsilon^{(1-\mu)/2})$ close to the primary strong canard in section Σ^1.*

Therefore all secondary canards converge to the primary strong canard in the singular limit $\varepsilon \to 0$. Furthermore the sizes of the intervals I'_j for $1 \leq j < k+1$ are $O(\varepsilon^{(1-\mu)/2})$ while the size of the interval I'_{k+1} is $O(1)$ because the two primary canards have distance $O(1)$ and therefore $I'_{k+1} = |\phi'_w - \phi'_k| = O(1)$.

Proposition 3.6 *Given system (3.1) and the section $\Sigma^1 : y = -\delta_1$. There exist $(k+1)$ sectors I'_j, $1 \leq j \leq k+1$, between the two primary canards where k is the number of secondary canards. The size of the sectors I'_j, $1 \leq j < k+1$, is $O(\varepsilon^{(1-\mu)/2})$ and the size of sector I'_{k+1} is $O(1)$.*

The qualitative picture is given in Figure 7.

As announced in the paragraph directly following Proposition 3.2 we are now in position to prove that $\phi_w(t)$ is the common boundary of S_a and S_r. More precisely, we prove the following:

Proposition 3.7 [26] *Consider (3.5) with $\varepsilon > 0$. As $\varepsilon \to 0$ the distance between the boundary of $S_{a,\sqrt{\varepsilon}}$ (resp. $S_{r,\sqrt{\varepsilon}}$) and the weak canard converges to 0.*

The idea is that any neighborhood of Φ_w in Σ^1 which does not contain Φ_s is contracted by the flow of (3.6) onto a neighborhood of Φ_w whose size converges to 0 as $\varepsilon \to 0$. This is intuitively clear from Figure 7. More precisely, we prove the following lemma.

Lemma 3.1 *Let I be a closed interval in Σ^1 including the intersection of Φ_w with Σ^1 in its interior but not the intersection of Φ_s with Σ^1. The map Π_2 restricted to I is a contraction with contraction rate $O(\varepsilon^{(1-\mu)/2\mu})$.*

Proof: The procedure is analogous as in the proof of Proposition 3.5 except in forward time. For this reason we need the linearization about P_w, which is given by:

$$\begin{pmatrix} \dot{x} \\ \dot{w} \end{pmatrix} = \begin{pmatrix} -\mu & 0 \\ * & -(1-\mu) \end{pmatrix} \begin{pmatrix} x \\ w \end{pmatrix}. \tag{3.8}$$

The forward transition time from Σ^1 to Σ^2 for trajectories not close to Φ_s is

$$T = -\frac{1}{\mu} \ln \sqrt{\varepsilon} + O(1).$$

We now argue, analogously as above, that the evolution of the variable transverse to Φ_w is given approximately by $\dot{w} = (\mu - 1)w$. The required result follows easily. □

Proof of Proposition 3.7: For definiteness let us assume that $p = (x, y, z)$ in $S_{a,\varepsilon} \cap \Sigma_1$ corresponds to the turning point in Σ^1. If p' is the image of p in Σ^2 under the flow of (3.1), then, by Lemma 3.1, p' is $O(\varepsilon^{(1-\mu)/2\mu})$ close to the weak canard (in the (x, w, y) coordinates). This means that p' in the $(\bar{x}, \bar{z}, \bar{y})$ coordinates must be $O(\varepsilon^{(1-\mu)/2\mu})$ close to the weak canard. As the weak canard converges to φ_w as $\varepsilon \to 0$, it follows that p' converges to the intersection point of φ_w with Σ^2. □

4 Global return mechanism and periodic orbits

In this section we state more precise versions of Theorem 2.1 and sketch the proofs. For precise proofs, based on the blow-up method, see our forthcoming article [5].

Recall Assumption 3 postulating the existence of a singular periodic orbit Γ passing through the singular funnel. We now make some additional assumptions on the critical manifold S and on Γ. For simplicity we assume that the critical manifold has S shaped cross-sections. More precisely we assume that S consists of two stable sheets S_1 and S_2, one unstable sheet S_r and two fold lines $\mathcal{F}_1$ and $\mathcal{F}_2$. Let $\mathcal{F}_1'$ (resp. $\mathcal{F}_2'$) be the projection of the fold line $\mathcal{F}_1$ (resp. $\mathcal{F}_2$) onto S_2 (resp. S_1) along the x-direction. We assume that $\mathcal{F}_1$ contains a folded node point and that there is a singular periodic orbit Γ as shown in Figure 8. We make the following assumption:

Assumption 3′ *The orbit Γ_g is defined as*

1. *A critical fiber connecting the folded node point to S_2,*
2. *a (non-singular) trajectory of (3.2) leading to a non-degenerate jump point*
3. *a critical fiber connecting the jump point to $\mathcal{F}_2'$.*
4. *The endpoint of Γ_g is either in the singular funnel or on the strong canard.*
5. *The reduced flow of (3.2) intersects $\mathcal{F}_j'$ transversely at the points $\Gamma \cap \mathcal{F}_j'$, j=1,2.*

Note that Assumption 3′ implies that a singular periodic orbit Γ as shown in Figure 8 exists.

Remark: The results stated below hold not only for an S-shaped return mechanism, but in much higher generality. In fact we only need non-degeneracy assumptions on the singular flow near Γ. However, in the context of this paper we prefer to use the clear and concrete picture given by Assumption 3′.

Remark: If Assumption 3′ does not hold then no MMO's can be expected. We explain this point under the additional assumption that the singular strong canard intersects $\mathcal{F}_2'$ transversely at one point p_s and all points in $\mathcal{F}_2'$ to on side

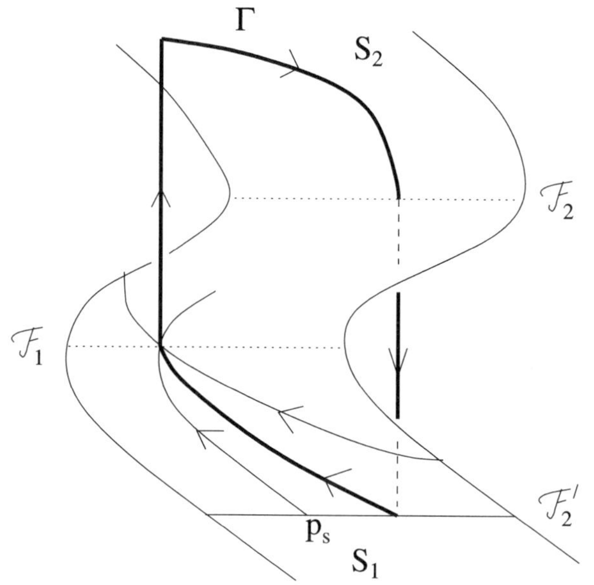

Figure 8 The cubic return mechanism and the singular periodic orbit Γ

of p_s are in the singular funnel (with no further loss of generality we assume that the singular funnel consists of the trajectories under (3.2) of the points in $\mathcal{F}_2'$ to the right of p_s, see Figure 8). Let p_1 be the endpoint of Γ_g and suppose that p_1 is to the left of p_s (see Figure 8). Then, by monotonicity of the reduced flow, the singular trajectory through p_1 must return to $\mathcal{F}_2'$ to the left of p_1, and so on. It follows that no MMO's can exist in this case.

We now present some positive results on the existence of simple MMO's. Recall the rotation sectors I_j', which were defined as intervals in $S_a \cap \Sigma_1$. We will now replace them by exponentially small neighborhoods of I_j' in Σ_1, keeping the same notation. We will make a standing assumption that μ is a a fixed constant between 0 and 1 and not at resonance. We will not make claims about the dependence of the length of I_j' on μ.

Theorem 4.1 *(Existence of 1^s MMO) Given system (2.1) under Assumptions 1, 2 and 3'. Assume also the endpoint of Γ_g is in the interior of the singular funnel region and not on the weak canard. Then, for sufficiently small $\varepsilon > 0$, there exists a stable periodic orbit passing through the sector of maximal rotation I_{k+1}'.*

Sketch of the proof: On the singular level the return map Π maps the entire singular funnel to a single point. It follows that for $\varepsilon > 0$ Π restricted to I_{k+1}' is a very strong contraction. Hence there exists a unique attracting periodic orbit passing through I_{k+1}'. □

Periodic orbits passing through submaximal sectors exist in the unfolding of a codimension one situation. More precisely, we assume that the system depends on a parameter β and variation of β has the effect of moving the the global return point in S_a. We have the following theorem.

Theorem 4.2 *Given system (2.1) under Assumptions 1, 2 and 3'. Assume also that for some $\beta = \beta_0$ the return point is on the strong canard and as β passes through β_0 the return point crosses the strong canard with non-zero speed. Then the following holds provided that ε is sufficiently small:*

For each j there is an interval $\tilde{I}_j$ with length of the order $O\left(\varepsilon^{\frac{1-\mu}{2}}\right)$ such that for each $\beta \in \tilde{I}_j$ there exists a periodic orbit passing through I'_j.

Sketch of the proof: The size of a submaximal sector in $\Sigma^- \cap S_{a,\varepsilon}$ is $O(\varepsilon^{(1-\mu)/2})$. In the vicinity of the fold the sectors are $O(1)$ in the $(\bar{x}, \bar{y}, \bar{z})$ variables, which suggests contraction by factor $O(\varepsilon^{\mu/2})$. From the vicinity of the fold back to Σ^- the sectors do not particularly contract or expand, except for trajectories near the canards. This suggests that the sector may be returned inside of itself and this will happen in the parameter range $\beta = O(\varepsilon^{(1-\mu)/2})$. The resulting map will be a uniform contraction with contraction rate $O(\varepsilon^{\mu/2})$. For the relevant interval of β there will be an attracting periodic orbit passing through the given sector. □

More complicated Farey sequences can occur if no sectors are returned inside themselves. In principle the only restriction is that the number of small oscillations is bounded, but our simulations indicate that only neighboring sectors can interact. The study of more complicated periodic orbits and possible chaos will be the subject of our future work.

Remark: As $\mu \to 0$, which corresponds to an approach to a folded saddle-node, the rate of contraction on the submaximal sectors decreases, in fact our estimate does not imply any contraction in the limit. Also the number of sectors grows indefinitely, while their size decreases to at most $O(\sqrt{\varepsilon})$. In fact most of the sectors in K_2 (the intervals I_j) must shrink to 0 in length as μ approaches 0. This suggests that near a folded saddle-node the dynamics becomes less dissipative, submaximal sectors play a greater role and trajectories visiting many different sectors are more likely to occur.

5 Simulations of MMO's

For our simulations we have used the system:

$$\begin{aligned} \varepsilon\dot{x} &= y + \frac{3}{2}x^2 - x^3 \\ \dot{y} &= -(\mu+1)x - z \\ \dot{z} &= -\frac{1}{3}(\mu - x + bx^2) \end{aligned} \tag{5.1}$$

Equation (5.1) is a 'van der Pol-like' toy model. There is a folded node point at the origin for any $\mu \in (0, 1)$ and mixed mode oscillations can be observed for a range of b values (b corresponds to β of the preceding section). An important concern was to pick μ sufficiently large relative to ε, in order to stay out of the folded saddle-node range, but not too large, so that the oscillations near the fold could be observed. A good choice of μ and ε turned out to be $\mu = 0.15$ and $\varepsilon = 0.001$. Note that $6 \leq \frac{1}{\mu} \leq 7$, so that our theory predicts the presence of three sectors of rotation, namely I'_1, I'_2 and I'_3.

The following figures show the time series of the periodic orbits found, corresponding to various cases of the theorems presented above. Note that the evolution of both x and z near the fold consists of oscillation and drift along the weak canard. To eliminate the drift we used $x+z$ as dependent variable. The region of existence of MMO's starts at $b \approx -0.9$ and 1^3 solutions exist for $b \approx -1.25$ or less. For our simulations we chose $b = -1.25$, which is close to the boundary of the sector. For smaller b's the first two rotations become extremely small so that they essentially cannot be seen.

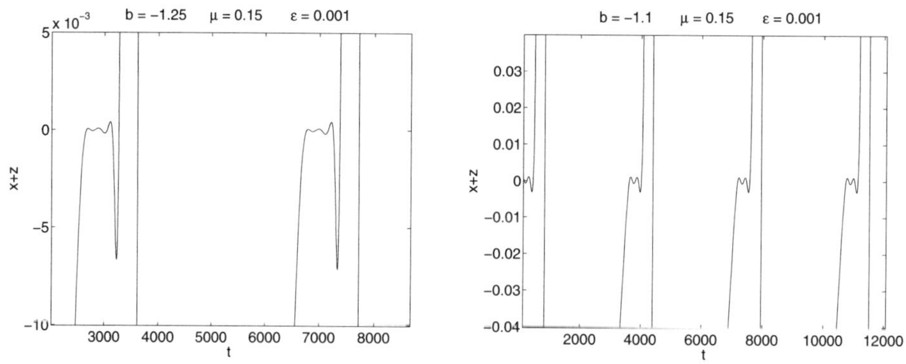

Figure 9 Left: Maximal sector solution MMO 1^3, sector I_3'; Right: MMO 1^2, periodic solution passing through I_2'

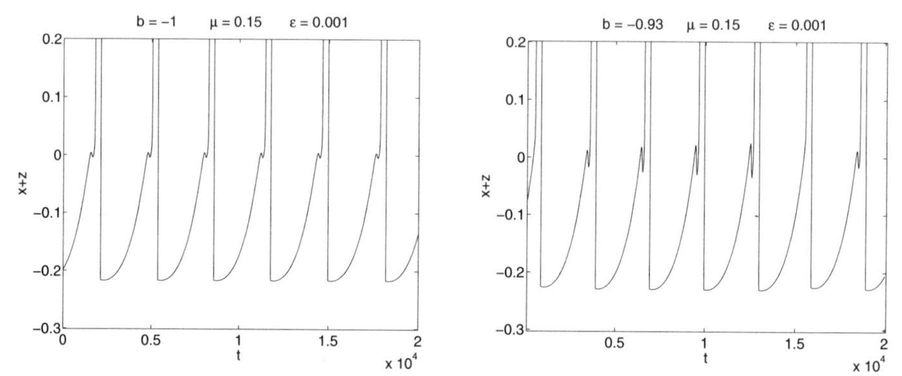

Figure 10 Left: MMO 1^1, periodic solution passing through I_1'; Right: Periodic solution MMO $1^1 1^1 1^1 1^1 1^0$ from the chaotic region

6 More fast dimensions

In this section we extend the theory presented above to the case of two slow and m fast dimensions, with m arbitrary. In this context a fold line is defined by the requirement that the m by m matrix of linearization of the fast equation in the fast directions should have a simple 0 eigenvalue. In this case there exists a local center manifold spanned by the slow variables and the critical direction of the fast variables. The local problem is governed by the flow on the center manifold, which is three dimensional, with one fast and two slow variables. Consequently it is possible to define the sectors of rotation and the funnel analogously as in the preceding sections, and to generalize the requirements on the global return. The results on MMO's in this generalized context are analogous as in the basic case.

A reduction theorem.
Given are equations of the form

$$\begin{aligned} \varepsilon \dot{x} &= f(x,y) \\ \dot{y} &= g(x,y) \qquad \text{with} \quad x \in \mathbb{R}^m,\ y \in \mathbb{R}^2 \end{aligned} \tag{6.1}$$

The above system has two slow and $m \geq 1$ fast variables. Let

$$S_0 = \{(x,y)\ :\ f(x,y) = 0\}$$

be the critical manifold. The following result shows how to obtain a reduction to systems with one fast and two slow variables.

Theorem 6.1 *Assume $(x_0, y_0) \in S_0$ is such that $D_x f(x_0, y_0)$ has a 0 eigenvalue with algebraic multiplicity 1 and all the other eigenvalues are in the left half plane. Then there exists a three dimensional center manifold M with the following properties:*
(1) *M is tangent to the three dimensional space spanned by the slow directions and $(v, 0, 0)$, where v is the nullvector of $D_x f(x_0, y_0)$*
(2) *M is exponentially attracting*
(3) *The vector field (6.1) reduced to M has a fold point of the type studied by Szmolyan and Wechselberger [26, 28]. More precisely (6.1) reduced to M has the form:*

$$\begin{aligned} \varepsilon \dot{\sigma} &= \tilde{f}(\sigma, y) \\ \dot{y} &= \tilde{g}(\sigma, y) \qquad \sigma \in \mathbb{R},\ y \in \mathbb{R}^2 \end{aligned} \tag{6.2}$$

with $\tilde{f}_\sigma(0, y_0) = 0$. The point $(\sigma, y) = (0, y_0)$ corresponds to (x_0, y_0).

Proof: The proof is completely analogous to the proof of the existence of a slow manifold. The center manifold M is non-unique, but any M is given as a graph as follows:

$$M := \{(x_0 + \sigma v + \phi(\sigma, y), y), \sigma \in (-\delta_0, \delta_0)\},$$

with δ_0 sufficiently small and $\phi = O(\sigma^2, (y - y_0)^2)$. The function ϕ is as smooth as f and g and if f and g are C^k then the k-jet of ϕ at $(0, 0, y_0)$ is uniquely determined. The functions $\tilde{f}$ and $\tilde{g}$ are defined as follows:

$$\begin{aligned} \tilde{f}(\sigma, y) &= w \cdot f(x_0 + \sigma v + \phi(\sigma, y), y) \\ \tilde{g}(\sigma, y) &= g(x_0 + \sigma v + \phi(\sigma, y), y) \end{aligned} \tag{6.3}$$

and w is the left nullvector of $D_x f(x_0, y_0)$, i.e. $w^T D_x f(x_0, y_0) = (0, 0)$. □

Theorem 6.2 *If, in addition to the assumptions of Theorem 6.1, we assume that*

$$w \cdot D_y f(x_0, y_0) g(x_0, y_0) = 0. \tag{6.4}$$

then the point (x_0, y_0) *is a folded singularity in the sense of* [26, 28].

Proof: According to [26, 28] system (6.2) has a folded singularity at $(0, y_0)$ if $\tilde{f}_y(0, y_0) \cdot \tilde{g}(0, y_0) = 0$. Clearly this condition is equivalent to (6.4). □

A point p_0 is now a folded singularity of (6.1) if it is contained in M and is a folded singularity of (6.2). Classification of folded singularities is analogous as in 3D.

Global return mechanism.
We assume that (x_0, y_0) is a folded node. The layer problem restricted to the invariant plane $y = y_0$ has saddle node point at x_0 with one center and $m-1$ stable direction. This condition guarantees that the flow on the center manifold for the layer problem restricted to the invariant plane $y = y_0$ approaches $x = x_0$ in one direction and moves away from $x = x_0$ in the other direction. In particular there is a trajectory going away from x_0. We refer to this trajectory as the *unstable critical fiber*. Now the generalization of the global return Γ_g is a sequence of trajectories of the layer flow and the reduced flow, starting with the unstable critical fiber and ending in the stable sheet of S_0 which contains the folded node point. The Assumptions 3 and/or 3' can now be easily generalized and theorems analogous to Theorem 4.2 and Theorem 4.1 hold.

7 Reduced flow and desingularization

In this section we derive a method for finding the desinguarized flow near the fold. The purpose of this is to develop an efficient way of finding folded singularities. Recall that in the basic case of two slow and one fast variables folded singularities are equilibrium points of the desingularized equation (3.3). Suppose for a problem with two slow and m fast variables a suitable generalization of the desingularized equation is available. One can then (possibly numerically) find equilibria and determine whether they correspond to folded singularities. The type of the folded singularity can be deduced from the properties of the linearization of the desingularized equation.

The reduced flow is given by the algebraic system:

$$\begin{aligned} 0 &= f(x, y) \\ \dot{y} &= g(x, y) \qquad \text{with} \quad x \in \mathbb{R}^m,\ y \in \mathbb{R}^2 \end{aligned} \tag{7.1}$$

In order to get a better understanding of the system (6.2) we would like to eliminate the constraint $f(x, y) = 0$. At a hyperbolic point of S_0 we can just solve for y as a function of x. Near the fold this is not possible, but one may be able to solve for either the fast variables x or for one slow variable and one fast variable. We will consider the following two cases:

(i) There exists $i, j \in \{1, \ldots, m\}$, $i \neq j$, such that $D_{(x_i, x_j)} f(x_0, y_0)$ has rank 2.
(ii) $D_{(x_i, y_j)} f(x_0, y_0)$ has rank 2, where $i \in \{1, \ldots, m\}$ and j equals 1 or 2.

The case $D_{(x_i,x_j)}f(x_0,y_0)$ has rank 2.

With no loss of generality we assume that $(i,j)=(1,2)$. We write $x=(\tilde{x},(x_3,x_4,\dots,x_m))$, with $\tilde{x}=(x_1,x_2)$. Let $(x_3,x_4,\dots,x_m,y)=\Phi(\tilde{x})$ denote the (possibly locally defined) solution of $f(x,y)=0$. We use $f(x,y)=0$ to derive the following identity:

$$0=\frac{d}{dt}(f(x,y))=D_xf(x,y)\dot{x}+D_yf(x,y)g(x,y)$$

or

$$D_xf(x,y)\dot{x}=-D_yf(x,y)g(x,y). \tag{7.2}$$

Let $C(x,y)$ denote the transpose of the co-factor matrix of $D_xf(x,y)$. We apply $C(x,y)$ to both sides of (7.2), getting

$$\det D_xf(x,y)\dot{x}=-C(x,y)D_yf(x,y)g(x,y). \tag{7.3}$$

Note that we need only two independent equations, namely the ones for $\dot{x}_1$ and $\dot{x}_2$. We have

$$\det D_xf(\tilde{x},\Phi(\tilde{x}))\dot{\tilde{x}}=-P_{(1,2)}C(\tilde{x},\Phi(\tilde{x}))D_yf(\tilde{x},\Phi(\tilde{x}))g(\tilde{x},\Phi(\tilde{x})), \tag{7.4}$$

where $P_{(1,2)}$ is the projection onto the first wo coordinates. For the important case of $m=2$ we have

$$C=\begin{pmatrix}\frac{\partial f_2}{\partial x_2} & -\frac{\partial f_1}{\partial x_2}\\ -\frac{\partial f_2}{\partial x_1} & \frac{\partial f_1}{\partial x_1}\end{pmatrix}$$

and

$$\det D_xf(x,\Phi(x))\dot{x}=-\begin{pmatrix}\frac{\partial f_2}{\partial x_2} & -\frac{\partial f_1}{\partial x_2}\\ -\frac{\partial f_2}{\partial x_1} & \frac{\partial f_1}{\partial x_1}\end{pmatrix}D_yf(x,\Phi(x))g(x,\Phi(x)) \tag{7.5}$$

Remark: If $(x(t),y(t))$ is a trajectory of (7.4) then it is the trajectory of (7.1).

We now rescale the time by $\det D_xf(x,\Phi(x))$ obtaining the reduced and desingularized equation:

$$\dot{\tilde{x}}=-P_{(1,2)}C(\tilde{x},\Phi(\tilde{x}))D_yf(\tilde{x},\Phi(\tilde{x}))g(\tilde{x},\Phi(\tilde{x})). \tag{7.6}$$

For $m=2$ we get

$$\dot{x}=-\begin{pmatrix}\frac{\partial f_2}{\partial x_2} & -\frac{\partial f_1}{\partial x_2}\\ -\frac{\partial f_2}{\partial x_1} & \frac{\partial f_1}{\partial x_1}\end{pmatrix}D_yf(x,\Phi(x))g(x,\Phi(x)) \tag{7.7}$$

Now we have the following proposition:

Proposition 7.1 *An equilibrium (x_0,y_0) of (7.6) is a folded singularity in the sense of Section 6 if $D_xf(x_0,y_0)$ has a zero eigenvalue of algebraic multiplicity* 1.

Proof: We assume for simplicity that $m=2$. In the general case the proof is similar.

If $D_xf(x_0,y_0)$ has a simple eigenvalue 0 then the vector

$$\left(\frac{\partial f_2}{\partial x_2}(x_0,y_0),-\frac{\partial f_1}{\partial x_2}(x_0,y_0)\right)$$

is either 0 or a left eigenvector of $D_xf(x_0,y_0)$. The same holds for

$$\left(-\frac{\partial f_2}{\partial x_1}(x_0,y_0),\frac{\partial f_1}{\partial x_1}(x_0,y_0)\right),$$

namely it is either 0 or a left eigenvector of $D_x f(x_0, y_0)$. It follows that (x_0, y_0) is an equilibrium of of (7.6) if and only if the condition (6.4) holds. □

The case $D_{(x_i,y_j)} f(x_0, y_0)$ has rank 2.

We assume for simplicity that $m = 2$. We also assume, with no loss of generality, that $i = j = 2$, i.e. $D_{(x_2,y_2)} f(x_0, y_0)$ is invertible. We can now find a smooth function Φ so that $(x_2, y_2) = \Phi(x_1, y_1)$. Further we proceed as in the case of two slow and one fast variable obtaining the reduced equation in the following form:

$$\begin{aligned} \dot{y}_1 &= g_1(x_1, y_1, \Phi(x_1, y_1)) \\ \frac{\det D_{(x_1,x_2)} f}{\det D_{(x_2,y_2)} f} \dot{x}_1 &= g_2(x_1, y_1, \Phi(x_1, y_1)) - \frac{\det D_(y_1, x_2) f}{\det D_{(x_2,y_2)} f} g_1(x_1, y_1, \Phi(x_1, y_1)). \end{aligned} \tag{7.8}$$

The desingularized flow is given by

$$\begin{aligned} \dot{y}_1 &= g_1(x_1, y_1, \Phi(x_1, y_1)) \frac{\det D_{(x_1,x_2)} f}{\det D_{(x_2,y_2)} f} \\ \dot{x}_1 &= g_2(x_1, y_1, \Phi(x_1, y_1)) - \frac{\det D_{(y_1,x_2)} f}{\det D_{(x_2,y_2)} f} g_1(x_1, y_1, \Phi(x_1, y_1)). \end{aligned} \tag{7.9}$$

Acknowledgement: The authors would like to express gratitude to Dr. Horacio Rotstein and Dr. Jonathan Rubin for carefully reading an earlier version of the manuscript.

References

[1] A. Arnéado, F. Argoul, J. Elezgaray and P. Richetti, Homoclinic chaos in chemical systems. *Physica D* **62** (1993) 134-168.

[2] E. Benoit, J-L. Callot, F. Diener, M. Diener, Chasse au canard, *Collect. Math.* **31** (1981), 37-119.

[3] K. Bold, C. Edwards, J. Guckenheimer, S. Guharay, K. A. Hoffman, J. Hubbard, Ricardo Oliva and W. Weckesser, The forced van der Pol equation II: canards in the reduced system. *SIAM Journal on Applied Dynamical Systems* **2** (2003), 570-608.

[4] K.M. Brucks and C. Tresser, A Farey Tree organization of locking regions for simple circle maps, *Proc. Amer. Math. Soc* **124** (1996), 637-647.

[5] M. Brøns, M. Krupa, M. Wechselberger, Folded nodes and mixed mode oscillations, in preparation (2006).

[6] C. T. Dickson, J. Magistretti, M. H. Shalinsky, E. Fransen, M. E. Hasselmo, A. Alonso, Properties and role of I(h) in the pacing of subthreshold oscillations in entorhinal cortex layer II neurons, *J Neurophysiol.* **83** (2000), 2562-79.

[7] Jonathan Drover, Jonathan Rubin, Jianzhong Su, and Bard Ermentrout. Analysis of a canard mechanism by which excitatory synaptic coupling can synchronize neurons at low firing frequencies. *SIAM J. Appl. Math.* **65** (2004), 65-92.

[8] I. R. Epstein, K. Showalter, Nonlinear chemical dynamics: oscillations, patterns, and chaos, *J. Phys. Chem.* **100** (1996), 13132.

[9] N. Fenichel, Geometric singular perturbation theory, *J. Differential Equations* **31** (1979), 53-98.

[10] J. Guckenheimer, K. A. Hoffman and W. Weckesser. The forced van der Pol equation I: the slow flow and its bifurcations. *SIAM Journal on Applied Dynamical Systems* **2** (2003), 1-35.

[11] J. Guckenheimer, M. Wechselberger, L.-S. Young, Chaotic attractors of relaxation oscillators, *Nonlinearity* **19** (2006), 701–720.

[12] C. K. R. T. Jones, Geometric singular perturbation theory, in *Dynamical Systems*, Springer Lecture Notes Math. **1609** (1995), 44-120.

[13] M. Koper, Bifurcations of mixed-mode oscillations in a three-variable autonomous van der Pol-Duffing model with a cross-shaped phase diagram, *Physica D* **80** (1995), 72-94.

[14] K. Krischner, M. Eiswirth, G. Ertl, Oscillatory CO oxidation on Pt(110): Modeling of temporal self-organization, *J. Chem. Phys.* **96** (1992), 9161-72.

[15] M. Krupa and P. Szmolyan. Extending geometric singular perturbation theory to non-hyperbolic points – fold and canard points in two dimensions. *SIAM. J. of Math. Anal.* **33** (2001), 286-314.

[16] R. Larter and C. G. Steinmetz. Chaos via mixed mode oscillations. *Phil. Trans. R. Soc. London* **A337** (1991), 241-298

[17] J. Maselko, H. L. Swinney, Complex periodic oscillations and Farey arithmetic in the Belousov–Zhabotinskii reaction, *J. Chem. Phys.* **85** (1986), 6430-41.

[18] G. Medvedev, J. Cisternas, Multimodal regimes in a compartmental model of the dopamine neuron, *Physica D* **194** (2004), 333-356.

[19] A. Milik, P. Szmolyan, H. Loeffelmann, E. Groeller, Geometry of mixed-mode oscillations in the 3d autocatalator, *Int. J. of Bifurcation and Chaos* **8** (1998), 505-519.

[20] C. A. Del Negro, C. G. Wilson, R. J. Butera, H. Rigatto, J. C. Smith, Periodicity, mixed-mode oscillations, and quasiperiodicity in a rhythm-generating neural network, *Biophys J.* **82** (2002), 206-14.

[21] H. G. Rotstein, T. Oppemann, J. A. White, N. Kopell. A reduced model for medial entorhinal cortex stellate cells: subthreshold oscillations, spiking and synchronization, submitted (2005).

[22] H. G. Rotstein, N. Kopell, A. Zhabotinsky, I. R. Epstein, A canard mechanism for localization in systems of globally coupled oscillators, *SIAM J. Appl. Math*, **63** (2003), 1998-2019.

[23] H. G. Rotstein, N. Kopell, A. Zhabotinsky, I. R. Epstein, Canard phenomenon and localization of oscillations in the Belousov-Zhabotinsky reaction with global feedback, *J. Chem. Phys.* **119** (2003), 8824-8832.

[24] H. G. Rotstein, R. Kuske, Localized and asynchronous patterns via canards in coupled calcium oscillators, accepted for publication in *Physica D* (2006).

[25] M. Schell, F. N. Albahadily, Mixed-mode oscillations in an electrochemical system. II. A periodic-chaotic sequence, *J. Chem. Phys.* **90** (1989), 822-828.

[26] P. Szmolyan, M. Wechselberger, Canards in $\mathbb{R}^3$, *J. Differential Equations* **177** (2001), 419-453.

[27] P. Szmolyan, M. Wechselberger, Relaxation Oscillations in $\mathbb{R}^3$, *J. Differential Equations* **200** (2004), 69-104.

[28] M. Wechselberger, Existence and Bifurcation of Canards in $\mathbb{R}^3$ in the Case of a Folded Node, *SIAM J. Appl. Dyn. Sys.* **4** (2005), 101-139.

[29] A. M. Zhabotinsky, Periodic kinetics of oxidation of malonic acid in solution, *Biofizika* **9** (1964), 306-311.

Fields Institute Communications
Volume **49**, 2006

Bioremediation of Waste in a Porous Medium

Changsheng Chen
Yahoo! Inc.
Sunnyvale, CA 94089 USA

John Chadam
Department of Mathematics
University of Pittsburgh
Pittsburgh, PA 15260 USA
chadam@pitt.edu

Abstract. A simplified model for bacterial remediation of wastes in a porous medium is proposed. The mathematical model consists of a coupled set of nonlinear partial differential equations and an ordinary differential equation. Basic existence, uniqueness and regularity of the system can be proved in three space dimensions, globally in time. The existence of traveling waves can also be established. In a physically relevant limit, a free boundary problem can be formally derived. In this sharp interface limit the shape stability of the moving reaction fronts can be studied using elementary bifurcation analysis. In agreement with industrial observations, these reaction fronts are stable in the early phases of remediation, becoming unstable as the concentration of bacteria becomes high. This destabilization of the reaction front could lead to undesirable fingering that could result in some of the waste being inaccessible to the bacteria, greatly reducing the effectiveness of this type of remediation process.

1 Introduction

In this section we describe the bioremdiation process, propose a mathematical model, and outline the main results.

Consider a region of soil that has been contaminated by an immobile waste product (e.g., coal tar). In general, soil harbours an enormous number of dormant bacterial species, some of which would eat away the contaminant if additional limiting nutrients (e.g., oxygen), required for this bacteria's growth, were also present.

2000 *Mathematics Subject Classification.* Primary 35L60, 35R35, 41A60, 65D30.

This work summarizes the Ph.D. dissertation research of the first author carried out under the supervision of John Chadam who participated in the workshop and presented the results.

Research of the second authors was partially supported by the National Science Foundation through grant DMS-9704567.

The current industrial strategy then is to flush the contaminated region with nutrients (and perhaps even particular strains of bacteria) so that the activated bacteria will remove the contaminant as part of their growth process, transforming the environmental waste into neutral compounds (e.g., water and carbon dioxide).

Bioremediation was first systematically studied by Borden and Bedient and their co-workers [1, 2]. A discussion of the techological and practical aspects of bioremediation can be found in [3]. The process has been modeled numerically in [4 - 7] among other workers. One of the first mathematical models for bioremediation was proposed by Odencrantz, Valocchi and Rittmann [8] and was elaborated upon and studied analytically by Oya and Valocchi [9] and by J. Xin and his co-workers [10 - 12]. These works model the biodegredation reaction using Monod kinetics

$$R = \frac{qcnm}{(K_m + m)(K_n + n)} \tag{1.1}$$

where q is the reaction rate coefficient, c is the concentration of bacteria, n is the concentration of nutrient, m is the concentration of the waste, and K_m, K_n are constants representing the half-maximum rate concentrations of m and n respectively. This model has the property (as observed in nature) that the concentration of the bacteria has an upper bound (see [10]). Moreover, by making a quasi-equilibrium assumption one can reduce the model to a two-component reaction-transport problem in which the concentration of the nutrient and bacteria move along together [10, 11].

Practitioners have noticed that the remediation process is effective if the concentrations of injected mobile species do not exceed critical values. At these higher levels they notice that unaltered regions begin to appear. In this note we will propose a new model with a nonlinear reaction-transport feedback mechanism that provides a possible explanation for this undesirable appearance of unaltered zones of contaminated soil.

Suppose that we model the soil as a homogeneous porous medium (i.e., the unaltered zones of interest will not arise as the result of heterogeneities but can only occur spontaneously due to self-organization caused by a destabilizing feedback mechanism). Let φ be the porosity of the porous medium, $m(x,t)$ the concentration of the solid, immobile waste material and $c(x,t)$ the aqueous concentration of injected material (nutrients, waste-specific bacteria). Here $t \in [0, \infty)$ is time and $x \in \Omega \subset \mathbf{R}^3$ where Ω is the spatial region of interest. The conservation of aqueous species is given by

$$\frac{\partial(\varphi c)}{\partial t} = \nabla \cdot (\varphi D \nabla c - \varphi c \mathbf{v}) - \rho \frac{\partial m}{\partial t} \tag{1.2}$$

where ρ is the (constant) density of the solid contaminant, D is the diffusion coefficient and $\mathbf{v}$ the fluid velocity given by Darcy's Law in terms of the pressure p;

$$\mathbf{v} = -\frac{\kappa}{\mu} \nabla p. \tag{1.3}$$

The appearance of the porosity φ in equation (1.2) reflects the fact that the fluid can only flow through the pores. In equation (1.3) κ is the permeability and μ is the viscosity. These equations are augmented by the conservation equation for the fluid

$$\frac{\partial(\rho_f \varphi)}{\partial t} = \nabla \cdot (\rho_f \varphi \mathbf{v}), \tag{1.4}$$

where ρ_f is the fluid density, and by an equation describing the nutrient/bacteria reaction with the waste;

$$\frac{\partial m}{\partial t} = -\ell mc(c_{eq} - c)_+. \tag{1.5}$$

This last equation models one of the simplest types of reaction that captures the dependence on the presence of the contaminant, the presence of the nutrient/bacteria along with a cutoff if the concentration of the nutrient/bacteria exceeds some equilibrium value c_{eq} (as in the preceding models [9 - 11]).

The destabilizing mechanism of interest here does not depend on the specific form of this reaction (see Section 4) so it suffices to present the discussion in terms of the simplest case (1.5). The destabilizing mechanism does, however, depend on the properties of the transport coefficients κ, μ and D. To begin, it is not unreasonable for them to be functions of the form $\kappa(c, \varphi), D(c, \varphi)$ and $\mu(c)$. While specific forms for these functions can be found in [13] only general properties will be required for the present discussion. Indeed, we can simplify further by taking the porosity φ to be a constant; i.e., the removal of the contaminant does not appreciably change the porosity of the soil (the case of porosity variations caused by changes in m was treated in previous work of one of the authors and his collaborators [14 - 17]). For the present discussion it is convenient to combine the permeability and viscosity into a single function

$$k(c) = \kappa(c)/\mu(c) \tag{1.6}$$

and to consider the two cases of $k(c)$ being an increasing or decreasing function of c. The former case might describe the situation when the concentration c is small allowing the bacteria to swim to areas of more food (waste) in addition to just being carried along by the flow. Another low concentration mechanism that might give rise to increasing k is that a developing thin coating of bacteria on an otherwise rough porous matrix would increasingly facilitate the flow of the nutrient/bacteria through the pore channels. At the other extreme, when the concentration c becomes large one can imagine the pores becoming blocked (i.e., $\kappa(c)$ decreasing) and/or the fluid becoming more viscous (i.e., $\mu(c)$ increasing), both leading to k being an decreasing function in this range of concentrations. One can likewise consider a variable diffusion coefficient, $D(c)$ (please see the remarks at the end of Section 4).

Returning to the problem (1.2) - (1.5) with φ and ρ_f constant, inserting (1.6), and transforming to the scaled variables $\tilde{c} = c/c_{eq}, \tilde{t} = \epsilon t$ with $\epsilon = \varphi c_{eq}/\rho << 1$, one obtains (dropping the tildes)

$$\epsilon c_t = \nabla \cdot (D(c)\nabla c + ck(c)\nabla p) - m_t \tag{1.7}$$

$$\nabla \cdot (k(c)\nabla p) = 0, \tag{1.8}$$

$$\epsilon m_t = -\ell mc(1 - c)_+. \tag{1.9}$$

For specificity we now take the spatial region to be the doubly infinite rectangular cylinder $\Omega = \mathbf{R} \times (0,1)^2$. We now impose the following boundary conditions at the inlet ($x = -\infty$) and the outlet ($x = +\infty$) of the channel: for $(y, z) \in (0,1)^2$ and all $t > 0$, at $x = -\infty$

$$c = c_i < 1, \tag{1.10a}$$

$$m = 0, \tag{1.10b}$$

$$\mathbf{v} = -k(c_i)p_x = A(t), \tag{1.10c}$$

and at $x = +\infty$

$$c = 1, \tag{1.10d}$$

$$m = M \text{ (constant)}, \tag{1.10e}$$

$$\mathbf{v} = -k(1)p_x = A(t). \tag{1.10f}$$

The first equation in (1.10) says there is a constant inlet concentration that is below the (scaled) equilibrium value. The second says that the remediation has progressed long enough for all of the contaminant to have been removed at the inlet and the reaction zone has now progressed to some point upstream. The third prescribes the inlet fluid velocity via Darcy's Law. Equation (1.10d) says that at the outlet the concentration has reached its equilibrium value after having passed through the entire reaction zone and (1.10e) that the contaminant is still in its original, unaltered state because the reaction zone has not yet reached this point. The last says that the outlet velocity must be the same as the inlet velocity because on the boundary of the constant cross-section of the channel $(0,1)^2$ we are assuming the no flow conditions

$$\frac{\partial c}{\partial n} = \frac{\partial p}{\partial n} = 0, \quad (y,z) \in (0,1)^2, \tag{1.10g}$$

where n is the normal. Finally, we complete the problem by imposing the initial conditions

$$c_i \leq c(x,y,z,0) = c_0(x,y,z) \leq 1, \tag{1.11a}$$

$$0 \leq m(x,y,z,0) = m_0(x,y,z) \leq M. \tag{1.11b}$$

The model equations (1.7 - 1.11) differ from those in [9 - 12] in several aspects. As mentioned earlier, our reaction (1.9) is simpler than the Monod kinetics but still captures its main features along with the physically important fact that the biomass (bacteria) has an upper bound for its concentration. The most significant difference is that in our model the diffusion and advection coefficients, $D(c)$ and $k(c)$, are concentration dependent. Although this leads to a nonlinear problem, we believe it is essential for the reaction-transport feedback mechanism that causes reaction fronts to destabilize and lead to unaltered contaminant zones as explained below. A second important difference is that we are interested in the shape stability analysis of planar fronts while the previous work examines one-dimensional fronts (planar [9 - 12] and spherical [18, 19]) and their stability within their own class. As a result our analysis must necessarily be higher dimensional while the previous work is exclusively one-dimensional. In order to explicitly derive traveling wave solutions for the $\epsilon \neq 0$ version of the problem in [9 - 12] the authors restrict their attention to the purely advective case (i.e., $D \equiv 0$), citing the common folklore that diffusion/dispersion has a stabilizing effect, including spreading of the reaction zones. In our work we retain the diffusive/dispersive effects (i.e., $D(c) \neq 0$) in order to examine more precisely the resulting stabilizing effects. In [9 - 12] an explicit traveling wave solution is available but for our more complicated model (1.7 - 1.11) it is not explicitly available when $D(c) \neq 0$ and $\epsilon \neq 0$. The planar traveling wave solution to problem (1.7 - 1.11) with $D(c) \neq 0$ can only be calculated explicitly in the sharp interface limit, $\epsilon \to 0$. As a consequence, we will study its shape stability within the class of sharp interface solutions, i.e., in the context of the moving free boundary problem obtained as the singular limit of problem (1.7 - 1.11) when $\epsilon \to 0$.

In the next section we summarize our rigorous results concerning the existence, uniqueness, regularity and asymptotic behaviour (as $\epsilon \to 0$) of the solutions to the model equations (1.7 - 11). A formal outline of how to obtain a sharp interface limit ($\epsilon \to 0$) to these equations is presented in Section 3. The resulting free boundary problem for the moving reaction interface is amenable to a shape stability analysis based on elementary bifurcation theory. In Section 4 we summarize the conclusions of this stability analysis showing that it confirms the previously cited observations of the practitioners. Specifically, we show that if D and k are constant (i.e., there is no feedback between reaction and flow through the concentration dependence of the transport coefficients) then planar reaction fronts remain stable. As a result, the contaminant is removed uniformily over the cross-section by the traveling, planar reaction front. In the "early" phase of the remediation, when the aqueous concentration, c, in the reaction zone is low and $k(c)$ is increasing, the reaction front is still stable. When the concentration c is large and pores are increasingly being plugged and the fluid viscosity is rising (i.e., $k(c)$) is a decreasing function) the planar reaction fronts lose stability over a bounded interval $(0, A)$ of wavelengths. This simultaneous destabilization of multiple modes can lead to fingering in the reaction front that, in turn, can result in contaminated regions being left unaltered, as observed in practice. For example, a single finger (or a small number of fingers) might develop through which the entire remediation material is channeled resulting in unaltered contaminated regions. Of course, higher order, nonlinear, effects can lead to patterned or highly destabilized reaction fronts that could revert to removing the contaminant over the entire cross-section. As in [17], where the thickness of the reaction zone was proven to be $O(\sqrt{\epsilon})$, one anticipates that here as well the analytical stability results to be summarized in Section 3 will provide an indication of what to expect in numerical studies of shape stability for the $\epsilon \neq 0$ equations (1.7 - 1.10) for small ϵ. While the stability criteria developed in Section 3 for the sharp interface problem might predict the onset of instabilities for equations (1.7 - 1.10), the higher order effects mentioned above are usually best studied numerically in the context of (1.7 - 1.10) where a reaction zone with finite thickness facilitates the numerical computations.

2 Rigorous results

Throughout this discussion we will assume that the transport coefficients are bounded, smooth functions satisfying

$$0 < a \leq D(s), k(s) \leq b \quad \text{for all } s \geq 0 \tag{2.1}$$

This restricts the porous matrix to never being completely plugged nor completely removed.

Theorem 1 *If $c_0 - c_i \in L^2$(near $x = -\infty$), $m_0 \in L^1$(near $x = -\infty$), $1 - c_0 \in L^2$(near $x = +\infty$) and $M - m_0 \in L^1$(near $x = +\infty$) then equations (1.6 - 11) have a global (in time) weak solution.*

The proof involves proving existence for the cutoff problem in $x \in [-N, N]$ by using Schauder's Fixed Point Theorem for our nonlinear parabolic-elliptic-ordinary system of differential equations. By obtaining appropriate a priori estimates one can take the $N \to \infty$ limit. The proof models that in [15 - 17] (with porosity dependent $D(\varphi), \kappa(\varphi)$) but, here, extends to three space dimensions whereas the original, porosity dependent, case has only been solved in two dimensions (i.e., in $(-\infty, \infty) \times (0, 1)$). In the variable porosity case [15 - 17] equation (1.4) leads to an inhomogeneous equation (1.8) which in turn presents difficulties in obtaining adequate estimates on the pressure gradient to prove existence and uniqueness in space dimensions $n > 2$ [17].

Uniqueness can be proved in the present case in three dimensions only if we assume a bound on the concentration and pressure gradients. A proof of the bound appearing in the hypothesis of the next theorem seems difficult because of the unboundedness of the spatial region.

Theorem 2 *If* $\sup_\Omega(|\nabla c| + |\nabla p|) \leq f(t)$ *for some increasing continuous function* f*, then the solution is unique.*

For sufficiently smooth data, regularity of the solution, including proof that it is a classical solution, can be obtained by adapting standard parabolic theory (sections 7, 8, Chapter II of reference [20]) to the present situation along the lines of [17].

We now direct our attention to finding traveling wave solutions which are relevant to studying the shape stability of the reaction zone. In particular, we specialize to the one space dimensional (planar) version of the problem and outline the existence of a constant velocity, traveling wave solution for all $\epsilon > 0$ using dynamical system methods. In the limit as $\epsilon \to 0$ we can also obtain an explicit traveling wave solution. This will motivate our study in the next section of the moving free boundary version of the problem.

In terms of the similarity variable $\xi = x - vt$, planar solutions have the form $c(x, y, z, t) = c(\xi)$ (similarly for m and p). Denoting derivatives with respect to ξ by a prime, equations (1.7 - 1.9) become

$$-\epsilon v c' = (D(c)c' + ck(c)p')' + vm', \tag{2.2}$$

$$(k(c)p')' = 0, \tag{2.3}$$

$$-\epsilon v m' = -\ell m c(1-c)_+. \tag{2.4}$$

These equations are to be solved subject to the boundary (inlet and outlet) conditions

$$c(-\infty) = c_i > 0, \quad m(-\infty) = 0, \quad c(\infty) = 1 \quad m(\infty) = M. \tag{2.5}$$

and (e.g., integrating (2.3) from $x = -\infty$ to $x = +\infty$ or using (1.9c) and (1.9f))

$$v_i = -k(c_i)p'(-\infty) = -k(1)p'(+\infty) = v_0. \tag{2.6}$$

That is, the inlet velocity v_i (equivalently the inlet pressure gradient, $p'(-\infty)$) are given and the outlet values are determined from (2.6). Integrating (2.2) from $\xi = -\infty$ to $\xi = +\infty$ and using (2.6) one obtains the speed of the travelling wave solution

$$v = v_i(1 - c_i)(M + \epsilon(1 - c_i))^{-1}, \tag{2.6a}$$

or in the original, unscaled variables $t_{orig} = t/\epsilon$ with $\epsilon = \varphi c_{eq}/\rho$ and $c = c_{orig}/c_{eq}$

$$V = \frac{\varphi}{\rho} v_i(c_{eq} - c_i)(M + \frac{\varphi}{\rho}(c_{eq} - c_i))^{-1}. \tag{2.6b}$$

This explicitly shows that in the original coordinates, the velocity of the front is much slower (since $\varphi << \rho$) than the characteristic fluid velocity v_i. Similarily, integrating from $\xi = -\infty$ to an arbitrary ξ one obtains

$$c' = \frac{1}{D(c)}\{(v_i - \epsilon v)(c - c_i) - vm\} \tag{2.7}$$

which, with equation (2.4), written as

$$m' = \frac{1}{\epsilon v} m c(1-c)_+, \tag{2.8}$$

is the system to be studied. The two relevant fixed points $(c_i, 0)$ and $(1, M)$ are respectively an unstable node and a saddle point. Standard phase plane analysis in the region $Q = \{(c, m), 0 < m < M, c_i + \frac{vm}{v_i - \epsilon v} < c < 1\}$ shows that an orbit emanating from the stable manifold of $(1, M)$ tends, as $\xi \to -\infty$, to $(c_i, 0)$ thus establishing the existence of a solution to equations (2.7, 2.8). In summary, one has (up to translations),

Theorem 3 *There is a unique traveling wave solution to problem (2.7, 2.8).*

The previous results for $\epsilon \neq 0$ depend on the smoothness and special form of the reaction. Theorems 1 - 3 are theoretical in nature with little about the solutions being explicitly calculable. When $\epsilon \to 0$, the reaction zone collapses to an interface and one expects that at this stage only the asymptotic features/consequences of the reaction are relevant rather than its specific form. Moreover, when $\epsilon \to 0$ the above solutions $(c_\epsilon(\xi), m_\epsilon(\xi))$ of problem (2.7, 2.8) have a limiting form which can be explicitly calculated and hence its stability can be studied analytically. The proof of the following Theorem is adapted from that for the porosity dependent case [15].

Theorem 4 *Up to translations,* $\lim_{\epsilon \to 0}(c_\epsilon(\xi), m_\epsilon(\xi))$ *exists and is equal to* $(\overline{c}(\xi), \overline{m}(\xi))$ *where*

$$\overline{m}(\xi) = \begin{cases} 0 & \xi < 0 \\ M & \xi > 0. \end{cases} \tag{2.9}$$

and

$$\overline{c}(\xi) = \begin{cases} c_i + (1 - c_i)e^{v_i \xi / D_i} & \xi < 0, \\ 1 & \xi > 0. \end{cases} \tag{2.10}$$

Here $v_i = -\kappa(c_i)p'(-\infty)$ *and* $D_i = D(c_i)$ *are the inlet velocity and diffusion coefficient.*

From an industrial point of view, this is the ideal solution - a planar reaction front moving along with a velocity which is much slower than the fluid velocity, eating away all the contaminant as it advances.

3 Free boundary problem

In the previous section we outlined the existence of solutions to problem (1.7 - 1.11) when $\epsilon \neq 0$, including a planar traveling wave solution. We also obtained an explicit sharp interface planar traveling wave solution (2.9 - 2.10) when $\epsilon \to 0$. In this section we begin the study of the shape stability of this solution within the class of sharp interface solutions. Specifically, we shall first obtain the sharp interface ($\epsilon \to 0$) version of equations (1.7 - 1.11) as a moving free boundary problem. It will be obvious that $(\overline{c}(\xi), \overline{m}(\xi))$ given by (2.9, 2.10) is a solution of these equations. In the following Section 4 we shall outline the shape stability analysis that confirms our observations presented at the end of the first section.

Formally, letting $\epsilon \to 0$ in equation (1.9), the reaction term on the right hand side vanishes when either $m = 0$ or $c = 1$ since $c \geq c_i > 0$. We will denote the interface between these two regions by $x = R(y, z, t)$ - the unknown moving free boundary. In the discussion to follow we will restrict our presentation to the two-dimensional case (i.e., all functions are z-independent). The same arguments and conclusions apply to the three-dimensional case solely at the expense of more complicated notation (e.g., wave numbers ℓ^2 in Section 4 are replaced by $\ell^2 + k^2$ for rectangular cross-sections).

Following standard asymptotic procedures, we expand all functions (c, m, p) in these two outer regions separated by the reaction interface as follows:

$$f^{outer} = f_0(x, y, t) + \epsilon^{1/2} f_1(x, y, t) + o(\epsilon^{1/2}). \tag{3.1a}$$

The outer solution (dropping the subscript zero on the leading order term) satisfies the equations:

$$\left.\begin{array}{l} \nabla \cdot (D(c)\nabla c + ck(c)\nabla p) = 0 \\ \nabla \cdot (k(c)\nabla p) = 0 \\ m \equiv 0 \end{array}\right\} \text{ in } x < R(y,t) \tag{3.2, 3.3, 3.4}$$

and

$$\left.\begin{array}{l} c \equiv 1 \\ \Delta p = 0 \\ m \equiv M \end{array}\right\} \text{ in } x > R(y,t). \tag{3.5, 3.6, 3.7}$$

Matching these outer solutions across the interface is accomplished by "magnifying" the thin reaction zone, using inner functions to study the relevant internal structure and matching the inner solutions with the outer solutions. Specifically, letting σ denote the upstream normal coordinate to the (unknown) interface $x = R(y,t)$, and τ the tangent coordinate, then with $\eta = \sigma/\sqrt{\epsilon}$ denoting the dilated coordinate, the inner functions have the form

$$f^{inner} = \tilde{f}_0(\eta, \tau, t) + \epsilon^{1/2} \tilde{f}_1(\eta, \tau, t) + o(\epsilon^{1/2}) \tag{3.1b}$$

Substituting these into equations (1.7 - 1.10), solving for the top order terms and imposing the usual matching conditions

$$\lim_{\eta \to \pm\infty} \tilde{f}_0(\eta, \tau, t) = \lim_{x \to R(y,t) \pm 0} f_0(x, y, t) \tag{3.1c}$$

provides the following "jump" conditions at the interface $x = R(y,t)$

$$c_- = c_+ (\equiv 1) \tag{3.8a}$$

$$p_- = p_+ \tag{3.8b}$$

$$\left(\frac{\partial p}{\partial n}\right)_- = \left(\frac{\partial p}{\partial n}\right)_+ \tag{3.8c}$$

$$D(c)\left(\frac{\partial c}{\partial n}\right)_- = M\ V_n \tag{3.8d}$$

where subscripts $-$ and $+$ indicate values downstream and upstream of the front, n is the normal in the direction of σ and V_n is the normal velocity of the front. Evidently, only the concentration gradient and the waste concentration are discontinuous across the front (since $\left(\frac{\partial c}{\partial n}\right)_+ = 0$ because $c_+ \equiv 1$ and $m_- = 0, m_+ = M$.) The motion of the interface is driven through condition (3.8d) containing these discontinuities. Equations (3.2 - 3.8) provide the sharp interface (moving free boundary) version of problem (1.7 - 1.11).

Recall that in Section 2 we outlined the proof of the existence of planar traveling wave solutions to equations (1.7 - 1.11) for all $\epsilon > 0$. In addition, we pointed that when $\epsilon \to 0$ in this class of planar solutions we obtained the solution (2.9, 2.10)

$$\overline{m}(x,t) = \begin{cases} 0 & x < vt \\ M & x > vt. \end{cases} \tag{3.9a}$$

$$\overline{c}(x,t) = \begin{cases} c_i + (1-c_i)\exp(v_i/D_i)(x-vt) & x < vt \\ 1 & x > vt. \end{cases} \tag{3.9b}$$

$$\overline{p}(x,t) = \begin{cases} p_0 - \int_0^{x-vt} \frac{v_i}{k(\overline{c}(s))}\, ds & x < vt \\ p_0 - v_i(x-vt)_{k_i} & x > vt. \end{cases} \tag{3.9c}$$

where p_0 is an arbitrary constant (pressure at the interface) and v is the unique travelling wave velocity obtained in (2.6a) with $\epsilon = 0$;

$$v = v_i(1-c_i)/M \tag{3.9d}$$

The form of the pressure solution is obtained by integrating (2.3) with (3.9b) substituted. It is a straightforward calculation to check that (3.9) is a planar solution to the free boundary problem (3.2 - 3.8). Indeed it can be shown to be the unique (up to translations) planar solution. In the next section, we will investigate its stability within the class of all sharp interface ($\epsilon = 0$) solutions; i.e., within the class of solutions to the free boundary problem (3.2 - 3.8).

4 Shape stability analysis

For notational convenience we let the size of the transverse dimension of the aquifer be π and translate the moving front to the origin by the change of variables $x' = x - vt$ (and drop the prime). One can then study the stability of small perturbations of the planar front (3.9) of the form

$$c(x) = \overline{c}(x) + \delta e^{\sigma t} c_1(x) \cos \ell y \tag{4.1a}$$

$$p(x) = \overline{p}(x) + \delta e^{\sigma t} p_1(x) \cos \ell y \tag{4.1b}$$

$$R(y,t) = 0 \quad + \delta e^{\sigma t} \cos \ell y. \tag{4.1c}$$

where δ is a small parameter, $\ell = 1, 2, \cdots$ is the wave number of the shape disturbance and $\sigma = \sigma(\ell)$ is the spectrum of the linearized problem. With the notation $D = D(\overline{c}), k = k(\overline{c}), D' = D'(\overline{c})$, etc., the linearized problem for c_1, p_1 can be written as: in $x < 0$

$$Dc_1'' + (2D'\overline{c}' + k\overline{p}')c_1' + (D''(\overline{c}')^2 - \ell^2 D + D'\overline{c}'')c_1 + k'\overline{c}'\overline{p}'c_1 + \ell^2 D\overline{c}' + k\overline{c}'\overline{p}_1' = 0, \tag{4.2a}$$

$$k(p_1'' - \ell^2 p_1 + \ell^2 \overline{p}') + (k\overline{p}'' + k''\overline{c}'\overline{p}')c_1 + k'\overline{p}'c_1' + k'\overline{c}'p_1' = 0 \tag{4.2b}$$

and in $x > 0$

$$c_1 \equiv 0 \tag{4.3a}$$

$$p_1'' - \ell^2 p_1 + \ell^2 \overline{p}' = 0, \tag{4.3b}$$

along with the asymptotic conditions

$$c_1 \to 0, \quad p_1' \to 0 \quad \text{as } x \to -\infty \tag{4.4a}$$

$$p_1' \to 0 \quad \text{as } x \to +\infty \tag{4.4b}$$

and the conditions on the interface $x = 0$

$$c_{1-} = 0 \tag{4.5a}$$

$$p_{1-} = p_{1+} \tag{4.5b}$$

$$p'_{1-} = p'_{1+} \tag{4.5c}$$

$$D(1)c'_{1-} = \sigma M. \tag{4.5d}$$

Equations (4.2 - 4.5) cannot be solved explicitly for arbitrary $D(c)$ and $k(c)$ and hence an expression for the spectrum is not available in general. Instead we summarize the stability results for a few illustrative cases that support our observations and objectives mentioned in the first section.

Example 1. $D(c) \equiv 1 \equiv k(c)$. This is the prototype of the case of constant (concentration independent) transport coefficients whose importance for this discussion is that all interaction between the reaction and the transport (through the concentration c) is precluded. Direct calculation shows that

$$p'_1(x) = 0, \quad -\infty < x < \infty \tag{4.6a}$$

and in $x < 0$

$$c_1(x) = v_i(1-c_i)e^{v_i x}\left(1 - \exp\left(\frac{\sqrt{v_i^2 + 4\ell^2} - v_i}{2}x\right)\right). \tag{4.6b}$$

Substituting (4.6b) into the interface condition (4.5d) gives for the spectrum of the ℓth mode

$$\sigma(\ell) = \frac{v_i(1-c_i)}{M}\,(v_i - (v_i + \sqrt{v_i^2 + 4\ell^2})/2) \tag{4.7}$$

which is clearly negative for each mode $\ell = 1, 2, \cdots$ and hence the planar front (3.9) is stable. Thus when there is no possibility of feedback between the reaction and the transport, the planar remediation front is stable to heterogeneous perturbations and moves through the aquifer, uniformly removing the contaminant. This is reminiscent of the (stable) Stefan problem modeling the melting of ice.

Example 2. $D(c) \equiv 1, k(c) = e^{ac}$ with $a > 0$. This corresponds to "early stage" remediation when $k(c)$ is increasing with c and, on phenomenological grounds, we expect the remediation front to remain stable. Following [21], we write equations (4.2) and (4.3) in terms of the horizontal velocity in $x < 0$

$$u = -k(c)p_x = -k(\bar{c})\bar{p}' + \delta u_1 e^{\sigma t}\cos \ell y \tag{4.8}$$

and eliminating c_1, one obtains for u_1 the problem

$$\left(\frac{d^2}{dx^2} - \ell^2 - v_i\frac{d}{dx}\right)\left(\frac{d^2}{dx^2} - \ell^2\right)u_1 - a(1-c_i)v_i e^{v_i x}\left(\frac{d}{dx} + v_i\right)\left(\frac{d^2}{dx^2} - \ell^2\right)u_1 = 0 \tag{4.9}$$

along with the boundary conditions as $x \to -\infty$

$$u_1 \to 0, \quad u''_1 \to 0 \tag{4.10a}$$

and on the interface $x = 0$

$$u'_1 = -\ell u_1, \; u''_1 = a\ell(1-c_i)v_i(u_1 - \ell v_i) + \ell^2 u_1. \tag{4.10b}$$

Thus the simple change of allowing k to be c-dependent has led to a complicated boundary value problem (4.9, 10) for the stability analysis. The equations become a bit more transparent by making the change of variables

$$t = -a(1-c_i)e^{v_i x}, \quad y = \left(\frac{d^2}{dx^2} - \ell^2\right)u_1 \tag{4.11}$$

that results in y satisfying the Kummer equation

$$\frac{d^2y}{dt^2} + \frac{dy}{dt} + \left(\frac{1}{t} - \left(\frac{\ell}{v_i}\right)^2 \Big/ t^2\right) y = 0. \tag{4.12}$$

One finds that the solution can be written as

$$y(t) = (-t)^{\mu+1/2} e^{-t}\ M(\mu - 1/2,\ 1 + 2\mu, t) \tag{4.13}$$

where $\mu = ((\ell/v_i)^2 + 1/4)^{1/2}$ and M is a Kummer function. Solving for u_1 (in 4.11), for c_1 (from (4.2a)) and substituting into the boundary condition (4.5d) one obtains

$$\sigma(\ell) = \frac{(1-c_i)}{M\ell^2} \Big\{ \ell^2 v_i^2 + a_2 \frac{dy}{dt}(-a(1-c_i))$$

$$+ a_2 y(-a(1-c_i)) + a_2 a_1 (1 - v_i) \left[2v_i (a(1-c_i))^{\ell/v_i} \right] \Big\} \tag{4.14a}$$

where the constants a_1, a_2 are given by

$$a_1 = \int_{-a(1-c_i)}^{0} (-s)^{\ell/v_i} s^{-1} y(s) ds \tag{4.14b}$$

$$a_2 = \frac{2a(1-c_i)\ell^2 v_i^2}{a_1 [a(1-c_i)]^{-(\ell/v_i+1)} + y(-a(1-c_i))}. \tag{4.14c}$$

Using properties of Kummer functions [22] one can numerically evaluate $\sigma(\ell)$ in (4.14) and find that for every $\ell = 1, 2, \cdots, \sigma(\ell)$ is negative. This again implies that the planar front is stable.

Example 3. $D \equiv 1, k(c) = e^{ac}, a < 0$. This corresponds to "late stage" remediation when blockage begins to occur and we expect the remediation front to be unstable. The analysis outlined in Example 2 can be repeated and now the expression (4.14) with $a < 0$ gives an interval $(0, A)$ of values of the wave number ℓ for which $\sigma(\ell) > 0 (\sigma(0) = \sigma(A) = 0)$. The size of the unstable interval, A, depends in a monotonically increasing manner on the Peclet number, $v_i L/D_i$, where L is the traverse width of the channel and D_i is the inlet diffusion coefficient. Indeed, the Peclet number is the bifurcation parameter. Thus, depending on the size of the Peclet number, one might have a large number of unstable models that could combine to lead to fingering in the front. This in turn may lead to the phenomenon observed by practitioners that some regions are left unaltered by the remediation process.

Example 4. $D(c) = 1 + hc, k = e^{ac}$. This example is intended to examine the effect of a variable diffusion coefficient on the above stability results. Unfortunately, closed solutions are not available in this case nor for $D(c) = \exp(hc)$. All of the results are obtained from numerical simulations based on shooting methods for solving the boundary value problems. As a test we checked that the algorithms confirm the results of the previous three examples when $h = 0$. We find that $h > 0$ (i.e., $D(c)$ increases with c) enhances stability and $h < 0$ (i.e., $D(c)$ decreases with c) enhances instability. This agrees with the qualitative statements in [9 - 12] and with the earlier comment that the Peclet number is the bifurcation parameter (note that the diffusion appears in the denominator). On the other hand, for $|h| < 1$ our numerical experiments never show that a qualitative change from the $D \equiv 1$ case can be effected by a change in h. That is, the situation in Example 2 of stability for

$D \equiv 1, k(c) = e^{ac}, a > 0$, cannot lead to instability by taking $D(c) = 1 + hc, h < 0$. More specifically, in this case $\sigma(\ell)$ rises but remains negative for all ℓ. Similarly, in Example 3 the interval of instability for $D \equiv 1, k(c) = e^{ac}, a < 0$, shrinks but does not disappear by taking $D(c) = 1 + hc, h > 0$.

Finally, we mention that there is a wealth of information on experiments, modeling, numerical simulation and analytical treatments of various aspects of remediation in the proceedings of a recent IMA workshop on the subject [23].

References

[1] R.C. Borden and P.B. Bedient, *Transport of dissolved hydrocarbons influenced by oxygen-limited biodegradation, 1, Theoretical development* Water Res. Research, **22** (1986) pp. 1973-1982.

[2] R.C. Borden, P.B. Bedient, M.D. Lee, C.H. Ward and J.T. Wilson, *Transport of dissolved hydrocarbons influenced by oxygen-limited biodegradation, 2, Field application*, Water Res. Research, **22** (1986), pp. 1983-1990.

[3] National Research Council, *In Situ Remediation, When Does It Work?*, Water Science and Technology Board, National Academy Press, Washington, D.C., 1993.

[4] F.J. Molz, M.A. Widdowson and L.D. Benefield, *Simulation of microbial growth dynamics, coupled to nutrient and oxygen transport in porous media*, Water Res. Research, **22** (1986), pp. 1207-1216.

[5] K. MacQuarrie and E. Sudicky, *Simulation of biodegradable organic contaminants in groundwater*, Water Res. Research, **22** (1990), pp. 1207-1216.

[6] C. Chiang, C. Dawson and M. Wheeler, *Modeling of in-situ bioremediation of organic compounds in groundwater*, Transport in Porous Media, **6** (1991), pp. 667-702.

[7] T.R. Ginn, C.S. Simmons and B.D. Wood, *Stochastic-convective transport with nonlinear reaction: Biodegradation with microbial growth*, Water Res. Research, **31** (1995), pp. 2689-2700.

[8] J. Odencrantz, A. Valocchi and B. Rittmann, *Modeling the interaction of sorption and biodegradation on transport in groundwater in situ bioremediation systems*, in Proceedings of the 1993 Groundwater Modeling Conference, E. Poeter et al., eds., Golden, CO, 1993.

[9] S. Oya and A. Valocchi, *Characterization of traveling waves and analytical estimation of pollutant removal in one-dimensional subsurface bioremediation modeling*, Water Res. Research, **33** (1997), pp. 1117-1127.

[10] J. Xin and D. Zhang, *Statistical analysis of biodegradation fronts in one-dimensional heterogeneous porous media*, Advances in Water Resources, **22** (1998), pp. 103-116.

[11] R. Murray and J. Xin, *Existence of traveling waves in a biodegradation model for organic contaminants*, SIAM J. Math. Anal., **30** (1999), pp. 72-94.

[12] J.X. Xin and J.M. Hyman, *Stability, Relaxation, and Oscillation of Biodegradation Fronts*, SIAM J. Applied Math., **61**, No. 2, (2000), pp. 472-505.

[13] J. Bear, *Dynamics of Fluids in Porous Media*, American Elsevier, Amsterdam, 1972.

[14] J. Chadam, P. Ortoleva and A. Sen, *Reactive percolation instabilities*, IMA J. Appl. Math., **36** (1987), pp. 207-220.

[15] J. Chadam, X. Chen, E. Comparini and R. Ricci, *Traveling wave solutions of a reaction-infiltration problem and a related free boundary problem*, Euro. J. Appl. Math., **5** (1994), pp. 255-265.

[16] X. Chen and J. Chadam, *A reaction-infiltration problem I: Existence, uniqueness, regularity and singular limit in one space dimension*, Nonlin. Anal. Theory, Meths. and Applics., **27** (1996), pp. 463-491.

[17] X. Chen, R. Gianni, R. Ricci and J. Chadam, *A reaction-infiltration problem: Existence, uniqueness and regularity of solutions in two space dimensions*, Math. Models and Meths. in Appl. Sci., **5** (1995), pp. 599-618.

[18] S. Oya and A. Valocchi, *Analytical approximation of biodegradation rate for an in situ bioremediation of groundwater under ideal radial flow conditions*, J. Contam. Hydrology, **31** (1998), pp. 275-293.

[19] J. Xin, *Fluid Limit and Free Boundaries in Biodegradation*, in Free Boundary Problems: Theory and Applications, N. Kenmochi, ed., GAKUTO Internat. Ser. Math. Sci. Appl., **13** Gakkōtosho, Tokyo, pp. 429-440.

[20] O.A. Ladyzenskaja, V.A. Solonnikov and N.N. Ural'ceva, *Linear and Quasilinear Equations of Parabolic Type*, Translations of Mathematical Monographs, Vol. 23, Providence, R.I., American Mathematical Society, 1968.

[21] J. Chadam, A. Peirce and P. Ortoleva, *Stability of reactive flows in porous media: Coupled porosity and viscosity changes*, SIAM J. Appl. Math., **51** (1991), pp. 684-692.

[22] M. Abramowitz and I.A. Stegun, *Handbook of Mathematical Functions*, U.S. Government Printing Office, Washington, D.C., 1964.

[23] Resource Recovery, Confinement, and Remediation of Environmental Hazards, J. Chadam, A. Cunningham, R. Ewing, P. Ortoleva and M. Wheeler eds., Volume 131, IMA Volumes in Mathematics and its Applications, Springer-Verlag, New York, Berlin, Heidelberg, 2000.

Fields Institute Communications
Volume **49**, 2006

Bifurcations of Gyroscopic Systems Near a 0 : 1 Resonance

R. J. McDonald
Boeing Satellite Systems, Inc.
2260 E. Imperial Highway
El Segundo, CA 90245, USA
robert.mcdonald@boeing.com

N. Sri Namachchivaya
Department of Aerospace Engineering
University of Illinois at Urbana–Champaign
306 Talbot Laboratory, 104 South Wright Street
Urbana, IL 61801, USA
navam@uiuc.edu

Wayne Nagata
Department of Mathematics
University of British Columbia
1984 Mathematics Road
Vancouver, B.C., Canada V6T 1Z2
nagata@math.ubc.ca

Abstract. We study the local and global bifurcation behavior of autonomous and nonautonomous gyroscopic systems near a 0 : 1 resonance. As an example, we study the bifurcations of a rotating shaft, focusing on the lightly damped, nearly symmetric autonomous case, and the parametrically excited, subharmonic resonance case. One goal of the analysis is to understand how energy may be transferred from the high frequency mode to the low frequency mode in these gyroscopic systems. Regions where energy transfer may occur from high to low frequency modes are identified. Finally, using recently developed bifurcation methods, in the subharmonic resonance case we detect the presence of orbits which are homoclinic to certain invariant sets. In the dissipative case, we are able to identify conditions under which a generalized Šilnikov orbit would exist. In certain parameter regions, we prove that such orbits exist which are homoclinic to fixed points on the slow manifold, leading to chaotic dynamics in the system. These orbits provide the mechanism by which energy transfer between modes may occur.

2000 *Mathematics Subject Classification.* Primary 37J40; Secondary 74H60.
Key words and phrases. gyroscopic system, normal form, bifurcation .

1 Introduction

Gyroscopic systems are an essential component in many engineering systems. A dynamical system may be classified as a *gyroscopic system* if its equations of motion possess terms of the form $\gamma_{ij}(q)\dot{q}_j$, where the gyroscopic coefficients γ_{ij} are skew symmetric, i.e. $\gamma_{ij} = -\gamma_{ji}$, and the q's are generalized coordinates of the system. Gyroscopic terms frequently arise in rotating systems in which the configuration is expressed relative to the rotating body. These gyroscopic terms couple two or more coordinates in the system. Typical examples of gyroscopic systems include rotating shafts, pipes conveying fluids, and moving belts with vibrations, e.g. chain saws. For example, a rotating shaft is considered a gyroscopic system when it is viewed in a rotating coordinate system. The transverse vibrations of the shaft, when viewed in a moving coordinate frame, generate Coriolis, or gyroscopic forces.

A general form of the equations of motion for a nonlinear nonautonomous gyroscopic system is given by

$$M\ddot{\mathbf{q}} + 2G\dot{q} + K\mathbf{q} + \nabla U(\mathbf{q}) + \zeta\left[\Lambda\dot{\mathbf{q}} + F(\mathbf{q},\dot{\mathbf{q}})\right] + \eta(\mathbf{q},\dot{\mathbf{q}},t) = 0 \tag{1.1}$$

In this equation, the generalized coordinates are given by $\mathbf{q} \in \mathbb{R}^n$. Such systems of ordinary differential equations (ODEs) are often obtained by applying a discretization procedure to the partial differential equations (PDEs) of motion. In (1.1), M, G, K, and Λ are constant $n \times n$ matrices, U is a scalar potential, F is a nonlinear vector function, and η is a time-dependent, vector-valued function. The symmetric matrix M represents mass terms in the equations, and is generally positive definite. The matrix G is skew-symmetric $\left(G = -G^T\right)$, and represents gyroscopic terms often arising from a Coriolis force. The matrix K represents linear stiffness terms in the equations of motion arising from a potential. The parameter ζ measures the damping in the system, and the linear damping terms are contained in the matrix B. The vector function η, linear in the coordinates $\mathbf{q}$ and $\dot{\mathbf{q}}$, represents time-dependent forcing terms in the system. The potential function U contains nonlinear conservative terms, while the vector function F contains nonlinear dissipative terms. Gyroscopic forces do no work in the equations of motion, unlike dissipative forces, which always cause the system to lose energy.

We shall consider gyroscopic systems with small damping and forcing, so it is vital to understand the behavior of the eigenvalues of the linear conservative gyroscopic system. In the absence of damping and forcing, the system described by (1.1) can be rewritten as a Hamiltonian system

$$\dot{\mathbf{x}} = A\mathbf{x} + JD_{\mathbf{x}}H_1(\mathbf{x}) \tag{1.2}$$

where $\mathbf{x} = (\mathbf{q},\mathbf{p}) \in \mathbb{R}^{2n}$ represents the generalized coordinates and momenta of the system. The $2n \times 2n$ matrix A is given by

$$A = \begin{bmatrix} -G & I_n \\ -\left(K - G^2\right) & -G \end{bmatrix},$$

where I_n is the $n \times n$ identity matrix, while the nonlinear Hamiltonian term is given by $H_1(\mathbf{x}) = U(\mathbf{q})$. The matrix J, sometimes known as the symplectic matrix, is given by

$$J = \begin{bmatrix} O_n & I_n \\ -I_n & O_n \end{bmatrix}$$

where O_n is the $n \times n$ zero matrix.

The eigenvalues of Hamiltonian systems are symmetric about both the real and imaginary axes of the complex plane. For these systems to be stable then, all eigenvalues must lie on the imaginary axis. Krein [13] was the first to describe how eigenvalues move in generic, parameterized families of Hamiltonian systems as the parameters are varied. In the absence of symmetries, simple eigenvalues will remain on the imaginary axis, while coincident eigenvalues will split and leave the imaginary axis. Symmetries, such as spatial reflections or rotations, modify this behavior. In the presence of such symmetries, the eigenvalues will pass each other, rather than splitting, as system parameters are varied. Thus, it is important to identify any physical symmetries that exist in the problem.

For simplicity, we shall consider two degree of freedom gyroscopic systems. Such systems are often obtained by truncating the Galerkin approximation for a continuous system which models the physical problem. Thus, a stable, two degree of freedom gyroscopic system will have two pairs of eigenvalues on the imaginary axis. In the presence of symmetry, these eigenvalues will pass without splitting as system parameters are varied. However, if small symmetry-breaking is present, or if the system has no symmetry, then coincident eigenvalues will split, causing an instability in the system. We consider the motion of the eigenvalues of the system as a bifurcation parameter δ, is slowly increased. Initially, the system is stable, as all eigenvalues are on the imaginary axis. Near a critical point, one pair of eigenvalues approaches the origin. Near this point, the ratio of the two frequencies is very small, $\frac{\omega_1}{\omega_2} \ll 1$. As the parameter δ is increased further, the eigenvalue pair near the origin splits, and moves onto the real axis, causing an instability in the system. As δ is increased further, the eigenvalues change their direction, and meet at zero once again. Both of these critical points correspond to non-semisimple double zero eigenvalue cases. As δ is increased further, the system is re-stabilized, as the eigenvalue pair near zero remains on the imaginary axis. This re-stabilization phenomenon is due to the presence of gyroscopic terms. For further increases in the bifurcation parameter δ, the two pairs of eigenvalues may meet along the imaginary axis, depending upon the direction of motion of the second eigenvalue pair. At this point, known as the Hamiltonian Hopf bifurcation point, the eigenvalues split, with two each moving into the left hand and right hand plane.

When damping is present in the system, the motion of the eigenvalues is altered. Very small damping basically shifts each of the eigenvalues towards the left half plane a small amount. Thus, initially the eigenvalues are in the left half plane, and move parallel to the imaginary axis. The eigenvalues meet on the negative real axis very close to zero, and split. Soon afterwards, one of the eigenvalues crosses into the right half plane, resulting in a simple bifurcation for the system. When this pair of eigenvalues meets again, it is on the positive real axis, and the eigenvalues again move parallel to the imaginary axis. Thus, for the damped system, there is no re-stabilization of the system. The only critical point for the damped system occurs near the first coincidence of the eigenvalues on the negative real axis, close to the origin. Thus, the second double zero eigenvalue and the Hamiltonian Hopf bifurcation for the Hamiltonian system do not correspond to critical points for the damped system. For this reason, the first critical point of the Hamiltonian system is more important to study, since it is near to a critical point in the weakly damped system.

1.1 $0:1$ **resonance.** For damped systems, the only modes that will not be damped out are those modes which are directly forced or whose frequencies are in resonance with the excitation frequency. For parametric excitation of a mechanical system, two types of resonances can occur: a subharmonic resonance or a combination resonance. A mode is said to be in subharmonic resonance if the excitation frequency is twice the frequency of that mode. A combination resonance occurs when the excitation frequency is equal to the sum of several natural frequencies in the system.

A different type of resonance, internal resonance, can occur when two (or more) natural frequencies in the system satisfy a relation like

$$k_1\omega_1 + k_2\omega_2 = 0 \tag{1.3}$$

where k_1 and k_2 are integers. Several common resonances occur when $\omega_1 = \omega_2$ and $\omega_1 = 2\omega_2$. Although not a resonance in the classical engineering sense, an internal resonance can occur when a system has widely spaced frequencies such that

$$\frac{\omega_1}{\omega_2} \ll 1$$

In this case, (1.3) is satisfied for any value of k_1 if $\omega_1 = 0$ and $k_2 = 0$. This resonance may exist when a system has a very low frequency mode, in addition to higher frequency modes. For example, many flexible engineering structures have frequency relationships like this. In such cases, a parametric excitation of the high frequency, low amplitude mode would be coupled by the resonance to the low frequency mode, possibly creating high amplitudes in the low frequency mode. Several authors have studied this problem (see Nayfeh and Mook [23], and Nayfeh and Nayfeh [24]) and observed the energy transfer from high energy modes to low energy modes. The gyroscopic systems that we study near the double zero eigenvalue case have such a relationship between the frequencies. Thus, one of the goals of this research is to observe and explain how energy may be transferred from the high frequency mode to the low frequency mode (i.e.the mode corresponding to the eigenvalues near zero).

Note that in the shaft problem presented in the Appendix, for the non-rotating system, the natural frequencies are given by $\bar{\omega}_1^2 = \pi^4$ and $\bar{\omega}_2^2 = \frac{I_{zz}}{I_{yy}}\pi^4$, where I_{yy} and I_{zz} are lateral area moments of inertia of the shaft cross section in the two directions transverse to the shaft axis. For the rotating system, the natural frequencies are given by $\omega_1^2 = \pi^4 - \Omega^2$ and $\omega_2^2 = \frac{I_{zz}}{I_{yy}}\pi^4 - \Omega^2$. Thus, the rotation of the shaft has the effect of decreasing its stiffness. If the rotation rate becomes large enough, the effective stiffness will go to zero, and the rotating shaft will exhibit whirling motion.

We focus on the subharmonic resonance case, since the global bifurcation methods we use are not applicable to the combination resonance case. The existence of chaotic invariant sets near slow manifolds was first considered by Kovacic and Wiggins [11]. However, their formulation neglects one technical condition, which is filled in by Haller [10], who also provides a review of much of the work in this area, and describes the multi-pulse method in some detail.

There have been many applications of the Šilnikov and multi-pulse methods to study the global dynamics of engineering systems, and we mention a few here. The first application of the Šilnikov method was in the study of the NLS equations by Kovacic and Wiggins [11]. Feng and Sethna [5] found single pulse Šilnikov orbits for thin, parametrically excited plates. Kovacic and Wettergren [12] found single and multi-pulse methods in resonantly driven coupled pendula. Malhotra and Namachchivaya [14] found single pulse Šilnikov orbits for the forced motion of suspended elastic cables and shallow arches. The first application of the multi-pulse method was by Haller and Wiggins [9] to study the NLS equations. Haller [10] gives several example problems using this method, including a forced beam, resonant surface wave interactions, and a nonlinear vibration absorber. Malhotra et al. [15] located single and multi-pulse orbits in the transverse motion of a spinning disk, using both the Šilnikov and multi-pulse techniques.

2 Normal forms

In order to determine the local dynamics of equations of motion, it is very helpful to first put the equations into normal form. This involves making a finite number of near-identity coordinate changes that simplify the lower order terms in the Taylor series expansion of the equations of motion, up to some finite order. The coordinate changes modify higher order terms as well, but typically these have little or no qualitative effect on the dynamics predicted by the simplified lower order terms. For general systems, there is a systematic procedure to transform differential equations into normal forms. This procedure is described in many books, for example, see Guckenheimer and Holmes [6]. A similar procedure to obtain normal forms exists for Hamiltonian systems (for example, see Meyer and Hall [21]).

In this section we describe the results of linear stability analysis for the rotating shaft. We also describe the normal forms for the nonlinear systems we study. We begin with Hamiltonian normal forms, which correspond to the undamped system. We consider the undamped shaft under a constant axial load, which as an abbreviation we refer to subsequently as the "unforced" case. In addition to the general case of a shaft with an elliptical cross-section, we also consider the case of the "nearly symmetric" system, a shaft with a nearly circular cross-section. Both the unforced and "forced" systems are considered, where the latter case means the system is parametrically excited due to a periodic variation in the shaft rotation rate. Finally, we consider the additional terms introduced to the normal forms when damping is present.

2.1 The Hamiltonian system and linear stability. For simplicity we first consider the unforced undamped system, where the shaft rotates with a steady rate Ω_0. The equations of motion in this case can be written as Hamilton's equations for a time-independent Hamiltonian function. We study the linear part of Hamilton's equations to determine the stability of the undeflected state (trivial solution). Then we describe the result of using Hamiltonian normal form theory, which systematically uses *symplectic* coordinate changes (coordinate changes that preserve the Hamiltonian structure) to simplify the lower order terms in the Hamiltonian function, and therefore simplify the lower order terms of Hamilton's equations of motion.

For the unforced undamped system, the Lagrangian is given by

$$L(\mathbf{q}, \dot{\mathbf{q}}) = \frac{1}{2}\dot{\mathbf{q}}^T\dot{\mathbf{q}} + \dot{\mathbf{q}}^T G\mathbf{q} - \frac{1}{2}\mathbf{q}^T K\mathbf{q} - U(\mathbf{q}),$$

and the corresponding equation of motion is

$$\ddot{\mathbf{q}} + 2G\dot{\mathbf{q}} + K\mathbf{q} + \nabla_{\mathbf{q}}U(\mathbf{q}) = 0, \tag{2.1}$$

where the term $\nabla_{\mathbf{q}}U(\mathbf{q})$ contains all the nonlinear conservative terms. The generalized momentum vector is defined by

$$\mathbf{p} = \frac{\partial L}{\partial \dot{\mathbf{q}}} = \dot{\mathbf{q}} + G\mathbf{q},$$

and we put $\mathbf{x} = (\mathbf{q}, \mathbf{p}) = [\,q_1\ q_2\ p_1\ p_2\,]^T \in \mathbb{R}^4$. We can then write the equation of motion (2.1) in Hamiltonian form as

$$\dot{\mathbf{x}} = A\mathbf{x} + JD_{\mathbf{x}}H_1(\mathbf{x}), \tag{2.2}$$

where

$$A = \begin{bmatrix} 0 & \Omega_0 & 1 & 0 \\ -\Omega_0 & 0 & 0 & 1 \\ -\bar{\omega}_1^2 & 0 & 0 & \Omega_0 \\ 0 & -\bar{\omega}_2^2 & -\Omega_0 & 0 \end{bmatrix}, \tag{2.3}$$

and the nonlinear term is given by $JD_{\mathbf{x}}H_1(\mathbf{x})$, where

$$H_1(\mathbf{x}) = H_1(\mathbf{q}, \mathbf{p}) = U(\mathbf{q}). \tag{2.4}$$

The total Hamiltonian for (2.2) is given by

$$H(\mathbf{x}) = \frac{1}{2}\left(p_1^2 + p_2^2\right) + \Omega_0\left(q_2 p_1 - q_1 p_2\right) + \frac{1}{2}\left(\bar{\omega}_1^2 q_1^2 + \bar{\omega}_2^2 q_2^2\right) + H_1(\mathbf{x}), \tag{2.5}$$

and Hamilton's equations of motion (2.2) have the form $\dot{\mathbf{x}} = JD_{\mathbf{x}}H(\mathbf{x})$.

Due to the elliptical shape of the shaft cross-section, the system has a discrete $\mathbb{Z}_2$-symmetry generated by

$$(\mathbf{q}, \mathbf{p}) \mapsto (-\mathbf{q}, -\mathbf{p}).$$

In particular, this implies that the nonlinear conservative term $U(\mathbf{q})$ is odd in $\mathbf{q}$, and the Hamiltonian function $H(\mathbf{x})$ is even in $\mathbf{x}$.

We study the system when the rotation rate Ω_0 is is increased, while the other parameters are held fixed. Here we consider the effect on the linearized equation $\dot{\mathbf{x}} = A\mathbf{x}$ and the stability of the equilibrium solution $\mathbf{x} = 0$ which represents the undeflected shaft. The eigenvalues of the matrix A given by (2.3) are

$$\pm i\omega_j = \pm i\sqrt{\Omega_0^2 + \frac{1}{2}\left(\bar{\omega}_1^2 + \bar{\omega}_2^2\right) \mp \sqrt{2\Omega_0^2\left(\bar{\omega}_1^2 + \bar{\omega}_2^2\right) + \frac{1}{4}\left(\bar{\omega}_1^2 - \bar{\omega}_2^2\right)^2}}.$$

The motion of the eigenvalues as the parameter is increased is shown in Figure 1. One pair of eigenvalues for the shaft always moves away from the origin along the imaginary axis, as Ω_0 is increased. Thus, there is no possibility of this mode becoming unstable, or a Hamiltonian Hopf bifurcation occurring. (If the parameter being changed was the end load exerted on the shaft, then this pair of eigenvalues would move towards the origin as the magnitude of the force was slowly increased. However, in this paper we only consider variation of the rotation rate.) As Ω_0 is increased from 0, the other pair of eigenvalues moves towards the origin along the imaginary axis, and reaches the origin at the smaller of the two critical frequencies $\bar{\omega}_j$. We assume that $\bar{\omega}_1^2 < \bar{\omega}_2^2$. Therefore, a pair of eigenvalues reaches the origin at $\Omega_0 = \bar{\omega}_1$, and we have the following eigenvalues and kernel:

$$0\,,\ 0\,,\ \pm i\sqrt{3\bar{\omega}_1^2 + \bar{\omega}_2^2}\,; \qquad \ker(A) = \operatorname{span}\left[1, 0, 0, \bar{\omega}_1\right],$$

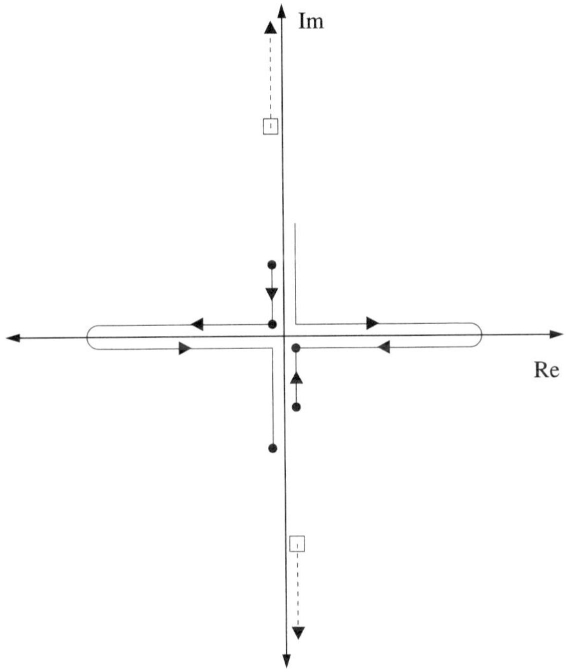

Figure 1 Eigenvalue motion for rotating shaft.

respectively. As Ω_0 is increased further, these eigenvalues split and move onto the real axis, turn around, and coalesce at the origin again at $\Omega_0 = \bar{\omega}_2$, where the eigenvalues and kernel are

$$0\,,\ 0\,,\ \pm i\sqrt{3\bar{\omega}_2^2 + \bar{\omega}_1^2}\,; \qquad \ker(A) = \text{span}\,[0, 1, -\bar{\omega}_2, 0]\,.$$

It is clear that these double zero eigenvalues correspond to non-semisimple Jordan forms. These eigenvalues then continue on the imaginary axis with further increases in Ω_0, moving away from the origin. For the shaft, the two modes considered represent the first mode of spatial deformation in each of two transverse directions.

2.2 Hamiltonian normal forms. In this subsection we describe the normal forms of the Hamiltonian function (2.5) and the corresponding equations of motion (2.2). We focus on the case $\Omega_0 = \bar{\omega}_1$, since this critical parameter value is where the undeflected shaft first becomes unstable. At $\Omega_0 = \bar{\omega}_1$, the eigenvalues of the matrix A are

$$0\,,\ 0\,,\ \pm i\Lambda_1,$$

where

$$\Lambda_1^2 = 3\bar{\omega}_1^2 + \bar{\omega}_2^2,$$

and the kernel is

$$\ker(A) = \text{span}\,[1, 0, 0, \bar{\omega}_1]\,,$$

We see that this double zero eigenvalue corresponds to a non-semisimple Jordan normal form for A.

The first step in our analysis is to calculate the linear normal form. By finding generalized eigenvectors, we form a complex symplectic matrix T so that the linear coordinate change

$$\mathbf{x} = T\mathbf{z},$$

where $\mathbf{z} = [\, z_1\ z_2\ \bar{z}_1\ \bar{z}_2 \,]^T \in \mathbb{C}^4$, puts the matrix A into complex normal form. The matrix T is given by

$$T = \left[\ u_1 + iv_1 \quad u_2 + iv_2 \quad u_1 - iv_1 \quad u_2 - iv_2\ \right] \tag{2.6}$$

where

$$u_1 = \begin{bmatrix} \frac{\sqrt{\bar{\omega}_2^2 - \bar{\omega}_1^2}}{\sqrt{2}\,\Lambda_1} \\ 0 \\ 0 \\ \frac{\bar{\omega}_1\sqrt{\bar{\omega}_2^2 - \bar{\omega}_1^2}}{\sqrt{2}\,\Lambda_1} \end{bmatrix}, \quad v_1 = \begin{bmatrix} -\frac{\sqrt{\bar{\omega}_2^2 - \bar{\omega}_1^2}}{\sqrt{2}\,\Lambda_1} \\ \frac{\sqrt{2}\,\bar{\omega}_1}{\Lambda_1\sqrt{\bar{\omega}_2^2 - \bar{\omega}_1^2}} \\ -\frac{\bar{\omega}_1^2 + \bar{\omega}_2^2}{\sqrt{2}\,\Lambda_1\sqrt{\bar{\omega}_2^2 - \bar{\omega}_1^2}} \\ -\frac{\bar{\omega}_1\sqrt{\bar{\omega}_2^2 - \bar{\omega}_1^2}}{\sqrt{2}\,\Lambda_1} \end{bmatrix},$$

$$u_2 = \begin{bmatrix} \frac{\bar{\omega}_1}{\Lambda_1^{3/2}} \\ -\frac{1}{2\sqrt{\Lambda_1}} \\ -\frac{\bar{\omega}_1}{2\sqrt{\Lambda_1}} \\ -\frac{\bar{\omega}_1^2 + \bar{\omega}_2^2}{2\Lambda_1^{3/2}} \end{bmatrix}, \quad v_2 = \begin{bmatrix} -\frac{\bar{\omega}_1}{\Lambda_1^{3/2}} \\ -\frac{1}{2\sqrt{\Lambda_1}} \\ -\frac{\bar{\omega}_1}{2\sqrt{\Lambda_1}} \\ \frac{\bar{\omega}_1^2 + \bar{\omega}_2^2}{2\Lambda_1^{3/2}} \end{bmatrix}.$$

The coordinate change takes the quadratic part of the Hamiltonian function into

$$i\left[\frac{1}{4}(z_1 - \bar{z}_1)^2 - \Lambda_1 z_2 \bar{z}_2\right],$$

with corresponding matrix

$$T^{-1}AT = \begin{bmatrix} -\frac{i}{2} & 0 & \frac{i}{2} & 0 \\ 0 & -i\Lambda_1 & 0 & 0 \\ -\frac{i}{2} & 0 & \frac{i}{2} & 0 \\ 0 & 0 & 0 & i\Lambda_1 \end{bmatrix}.$$

To study the system when the parameter Ω_0 is near the critical value, we put

$$\Omega_0 = \bar{\omega}_1 + \delta,$$

where δ is small. Then the matrix A becomes

$$A = \begin{bmatrix} 0 & \bar{\omega}_1 + \delta & 1 & 0 \\ -\bar{\omega}_1 - \delta & 0 & 0 & 1 \\ -\bar{\omega}_1^2 & 0 & 0 & \bar{\omega}_1 + \delta \\ 0 & -\bar{\omega}_2^2 & -\bar{\omega}_1 - \delta & 0 \end{bmatrix} \tag{2.7}$$

We can break this matrix up as

$$A = A_0 + \delta A_1$$

where

$$A_0 = \begin{bmatrix} 0 & \bar{\omega}_1 & 1 & 0 \\ -\bar{\omega}_1 & 0 & 0 & 1 \\ -\bar{\omega}_1^2 & 0 & 0 & \bar{\omega}_1 \\ 0 & -\bar{\omega}_2^2 & -\bar{\omega}_1 & 0 \end{bmatrix}, \qquad A_1 = \begin{bmatrix} 0 & 1 & 0 & 0 \\ -1 & 0 & 0 & 0 \\ 0 & 0 & 0 & 1 \\ 0 & 0 & -1 & 0 \end{bmatrix}.$$

We use the matrix T given in (2.6) to normalize A_0.

Hamiltonian normal form theory (see for example [21]) can now be used to simplify higher order terms in the Hamiltonian for the system, by introducing a finite number of symplectic near-identity coordinate changes. Using the $\mathbb{Z}_2$ reflection symmetry, straightforward calculations give the $\mathbb{Z}_2$-symmetric normal form of the Hamiltonian function up to fourth order:

$$\begin{aligned} K(\mathbf{z}) &= i\left[\frac{1}{4}(z_1-\bar{z}_1)^2+\frac{\beta_1\delta}{4}(z_1+\bar{z}_1)^2-(\Lambda_1-\beta_2\delta)\,z_2\bar{z}_2\right.\\ &\qquad \left.+\alpha_1(z_1+\bar{z}_1)^4+\alpha_2(z_1+\bar{z}_1)^2 z_2\bar{z}_2+\alpha_3 z_2^2\bar{z}_2^2\right], \end{aligned} \tag{2.8}$$

where the fourth order normal form coefficients α_j and β_j are real and can be determined from the equations of the rotating shaft (see the Appendix). In particular, we have

$$\alpha_1<0, \qquad \alpha_2<0, \qquad \alpha_3<0,$$

and

$$\beta_1>0, \qquad \beta_2<0.$$

The corresponding equations of motion are

$$\begin{aligned} \dot{z}_1 &= i\left[-\frac{1}{2}(z_1-\bar{z}_1)+\frac{\beta_1\delta}{2}(z_1+\bar{z}_1)+4\alpha_1(z_1+\bar{z}_1)^3+2\alpha_2(z_1+\bar{z}_1)z_2\bar{z}_2\right],\\ \dot{z}_2 &= i\left[-(\Lambda_1-\beta_2\delta)z_2+\alpha_2(z_1+\bar{z}_1)^2 z_2+2\alpha_3 z_2^2\bar{z}_2\right]. \end{aligned} \tag{2.9}$$

The nearly symmetric system. Here we consider the special case of the nearly symmetric sytem. If the shaft has a circular cross-section, the system has an S^1-symmetry, and in this case $\bar{\omega}_1=\bar{\omega}_2$. A shaft with an elliptical but noncircular cross-section has unequal frequencies $\bar{\omega}_1\neq\bar{\omega}_2$ and the system has $\mathbb{Z}_2$-symmetry, where $\mathbb{Z}_2\subset S^1$. Thus if the shaft has an elliptical but nearly circular cross-section, then the system may be treated as a small symmetry-breaking perturbation of the circular case, with $\bar{\omega}_1\approx\bar{\omega}_2$. A paper by Nagata and Namachchivaya [22] adopts this viewpoint, and we summarize their results.

We first describe the "perfectly symmetric" system, representing a rotating shaft with a circular cross-section. In this case, due to the circular symmetry, the equation of motion (2.2) is equivariant with respect to the action of the group S^1 of the rotations of the circle, given by

$$(\mathbf{q},\mathbf{p})\mapsto(R_\theta\mathbf{q},R_\theta\mathbf{p}),$$

for all $\theta\in S^1=\mathbb{R}/(2\pi\mathbb{Z})$, where R_θ is the 2×2 rotation matrix

$$R_\theta=\begin{bmatrix}\cos\theta & \sin\theta\\ -\sin\theta & \cos\theta\end{bmatrix}.$$

In particular the S^1-symmetry implies $\bar{\omega}_1=\bar{\omega}_2$, and the matrix A becomes

$$A=\begin{bmatrix}0 & \Omega_0 & 1 & 0\\ -\Omega_0 & 0 & 0 & 1\\ -\bar{\omega}^2 & 0 & 0 & \Omega_0\\ 0 & -\bar{\omega}^2 & -\Omega_0 & 0\end{bmatrix},$$

where $\bar{\omega}$ denotes the common value of $\bar{\omega}_1$ and $\bar{\omega}_2$. The eigenvalues in this case are given by

$$\pm i\,(\Omega_0+\bar{\omega}), \qquad \pm i\,(\Omega_0-\bar{\omega}).$$

For Ω_0 near 0, we have two pairs of imaginary eigenvalues, both near $\pm i\bar{\omega}$. As Ω_0 is then increased, one pair of eigenvalues moves towards the origin, while another pair moves away from the origin. When $\Omega_0 = \bar{\omega}$, the eigenvalues pass at zero, rather than split, and there are no further points at which eigenvalues are coincident. We see that when $\Omega_0 = \bar{\omega}$, the system has a double zero eigenvalue, with two other eigenvalues $\pm 2i\bar{\omega}$. However, the kernel of A at $\Omega_0 = \bar{\omega}$ is in this case two-dimensional, and hence this double zero eigenvalue is semisimple.

We now consider the normal form for the perfectly symmetric system. For Ω_0 near its critical value of $\bar{\omega}$, we introduce a bifurcation parameter δ near 0, by defining

$$\Omega_0 = \bar{\omega} + \delta.$$

Using this notation, the eigenvalues of A are $\pm i\delta$ and $\pm i(2\bar{\omega} + \delta)$. By finding the corresponding eigenvectors, one forms a complex symplectic matrix T such that the linear coordinate change

$$\mathbf{x} = T\mathbf{z}$$

transforms the quadratic part of the Hamiltonian function (2.5) into

$$i[\,-\delta z_1\bar{z}_1 + (2\bar{\omega} + \delta)z_2\bar{z}_2\,].$$

This coordinate change takes the matrix A into its complex normal form

$$T^{-1}AT = \begin{bmatrix} -i\delta & 0 & 0 & 0 \\ 0 & i(2\bar{\omega}+\delta) & 0 & 0 \\ 0 & 0 & i\delta & 0 \\ 0 & 0 & 0 & -i(2\bar{\omega}+\delta) \end{bmatrix},$$

which corresponds to the quadratic part of the transformed Hamiltonian.

As for the general case, Hamiltonian normal form theory can now be used to find higher order terms in the Hamiltonian normal form for the system. The S^1-symmetry is expressed in the new variables as equivariance under

$$(z_1, z_2) \mapsto (e^{-i\theta}z_1, e^{i\theta}z_2).$$

It is found that the S^1-symmetric normal form for the Hamiltonian (2.5), up to fourth order, has the form

$$K(\mathbf{z}) = i\left[-\delta z_1\bar{z}_1 + (2\bar{\omega}+\delta)z_2\bar{z}_2 + \frac{1}{2}A_0 z_1^2\bar{z}_1^2 + B_0 z_1\bar{z}_1 z_2\bar{z}_2 + \frac{1}{2}C_0 z_2^2\bar{z}_2^2\right], \quad (2.10)$$

where the Hamiltonian normal form coefficients A_0, B_0 and C_0 are real, and can be determined from the nonlinear conservative term $U(\mathbf{q})$ for the shaft (see the Appendix). The corresponding equations of motion are

$$\begin{aligned} \dot{z}_1 &= i[\,-\delta z_1 + A_0 z_1^2\bar{z}_1 + B_0 z_1 z_2\bar{z}_2\,], \\ \dot{z}_2 &= i[\,(2\bar{\omega}+\delta)z_2 + B_0 z_1\bar{z}_1 z_2 + C_0 z_2^2\bar{z}_2\,] \end{aligned}$$

Now we consider the nearly symmetric system, the rotating shaft with an elliptical but nearly circular cross-section, where the deviation from a perfectly circular cross-section is treated as small symmetry-breaking perturbation. Symmetry-breaking is introduced by putting $\bar{\omega}_1^2 = \bar{\omega}^2 - \tilde{\epsilon}$ and $\bar{\omega}_2^2 = \bar{\omega}^2 + \tilde{\epsilon}$ in (2.3), where $\tilde{\epsilon} \geq 0$ is a small symmetry-breaking parameter. When $\tilde{\epsilon} = 0$ we have a circular cross-section and the system has S^1-symmetry. The case $\tilde{\epsilon} > 0$ represents an elliptical cross-section, and the system retains the discrete $\mathbb{Z}_2$ reflection symmetry of the general case.

The eigenvalues of the matrix A are as described at the beginning of this section. As Ω_0 is increased from 0, one pair of eigenvalues on the imaginary axis move away from the origin. The other pair of eigenvalues move towards the origin until at $\Omega_0 = \bar{\omega}_1 = \sqrt{\bar{\omega}^2 - \tilde{\epsilon}}$ there is a non-semisimple double zero eigenvalue. Then as Ω_0 increases further, the eigenvalues split and move onto the real axis. At $\Omega_0 = \sqrt{4\bar{\omega}^4 - \tilde{\epsilon}^2}/(2\bar{\omega})$, these eigenvalues turn around and move back towards the origin along the real axis. At this "turnaround point", these eigenvalues are $\pm\tilde{\epsilon}/(2\bar{\omega})$. Thus, if $\tilde{\epsilon}$ is small, the eigenvalues will not have moved far onto the real axis before returning towards the origin. At $\Omega_0 = \bar{\omega}_2\sqrt{\bar{\omega}^2 + \tilde{\epsilon}}$, the eigenvalues have returned to zero, and there is once again a non-semisimple double zero eigenvalue. For further increases in Ω_0, both pairs of eigenvalues move away from the origin on the imaginary axis, with no further intersections.

To calculate the Hamiltonian normal form for the case when symmetry-breaking is small, Nagata and Namachchivaya [22] use the symmetric, semisimple linear normal form as an organizing center, but incorporate the non-semisimple normal forms for the non-symmetric case through an unfolding. The normal form for the nonlinear terms is then calculated with the semisimple linear normal form corresponding to the symmetric case. This makes the normal form calculations much easier, but still in effect incorporates the normal forms of the two non-semisimple double zero eigenvalue cases.

We introduce the bifurcation parameter δ, and a new symmetry-breaking parameter $\beta = \mathcal{O}(\tilde{\epsilon}) \geq 0$, by

$$\Omega_0 = \bar{\omega} + \delta, \qquad \tilde{\epsilon} = \beta\sqrt{4\bar{\omega}(\bar{\omega} + \delta) + \beta^2}.$$

The choice of β rather than $\tilde{\epsilon}$ as a symmetry-breaking parameter simplifies computations. The parameters δ and β are small, and are treated as being of the same order of smallness. The eigenvalues of the matrix A become $\pm\sqrt{\beta^2 - \delta^2}$ and $\pm i\sqrt{(2\bar{\omega} + \delta)^2 + \beta^2}$, and one forms a complex symplectic matrix

$$T = T(\delta, \beta) \tag{2.11}$$

such that the linear coordinate change $\mathbf{x} = T\mathbf{z}$ takes the quadratic part of the Hamiltonian function (2.5) into the normal form

$$i\left[\left(-\delta + \frac{\beta^2}{2\bar{\omega}}\right) z_1\bar{z}_1 + \sqrt{(2\bar{\omega} - \delta)^2 + \beta^2}\, z_2\bar{z}_2 + \frac{\tilde{\epsilon}}{4\bar{\omega}}(z_1^2 + \bar{z}_1^2)\right].$$

Since δ and β are small, it is sufficient to consider only S^1-symmetric terms at fourth order in the Hamiltonian normal form, and furthermore, expressions involving δ and β may be simplified by keeping only lower order terms. Thus, we obtain the Hamiltonian normal form

$$\begin{aligned} K(\mathbf{z}) &= i\left[-\delta z_1\bar{z}_1 + \left(2\bar{\omega} + \delta + \frac{\beta^2}{4\bar{\omega}}\right) z_2\bar{z}_2 + \frac{\beta}{2}(z_1^2 + \bar{z}_1^2)\right. \\ &\qquad \left. + \frac{1}{2}A_0 z_1^2\bar{z}_1^2 + B_0 z_1\bar{z}_1 z_2\bar{z}_2 + \frac{1}{2}C_0 z_2^2\bar{z}_2^2\right], \end{aligned} \tag{2.12}$$

and the corresponding equations of motion are

$$\begin{aligned} \dot{z}_1 &= i\left[-\delta z_1 + \beta\bar{z}_1 + A_0 z_1^2\bar{z}_1 + B_0 z_1 z_2\bar{z}_2\right], \\ \dot{z}_2 &= i\left[\left(2\bar{\omega} + \delta + \frac{\beta^2}{4\bar{\omega}}\right) z_2 + B_0 z_1\bar{z}_1 z_2 + C_0 z_2^2\bar{z}_2\right]. \end{aligned} \tag{2.13}$$

2.3 Normal forms for the forced system. Returning to the general case, in the Hamiltonian framework we can also consider the forced (i.e. parametrically excited) system, where the rotation rate of the shaft has the form

$$\Omega(t) = \Omega_0 + h \cos \nu t,$$

where Ω_0 and ν are positive $\mathcal{O}(1)$ constants, and $h > 0$ is a small parameter. In this case the matrix $G = G(t)$ is periodic in t, the equation of motion

$$\ddot{\mathbf{q}} + 2G(t)\dot{\mathbf{q}} + K\mathbf{q} + \nabla_{\mathbf{q}} U(\mathbf{q}) = 0, \tag{2.14}$$

and can be written in Hamiltonian form as

$$\dot{\mathbf{x}} = A\mathbf{x} + JD_{\mathbf{x}}H_1(\mathbf{x}) + h \cos \nu t \, P\mathbf{x},$$

where

$$A = \begin{bmatrix} 0 & \Omega_0 & 1 & 0 \\ -\Omega_0 & 0 & 0 & 1 \\ -\bar{\omega}_1^2 & 0 & 0 & \Omega_0 \\ 0 & -\bar{\omega}_2^2 & -\Omega_0 & 0 \end{bmatrix}, \qquad P = \begin{bmatrix} 0 & 1 & 0 & 0 \\ -1 & 0 & 0 & 0 \\ 0 & 0 & 0 & 1 \\ 0 & 0 & -1 & 0 \end{bmatrix}.$$

To study the system when the rotation rate is oscillating about a value near the critical rotation rate for the unforced system, we put

$$\Omega_0 = \bar{\omega}_1 + \delta.$$

Then we may write

$$A = A_0 + \delta A_1,$$

as above in the unforced system. We also detune the equations of motion, to study what happens when the forcing frequency ν is near, but not necessarily at, some critical forcing frequency ω_0 (which could be, for example, some integer multiple of the internal frequency Λ_1). To detune the system, we put

$$\nu = \omega_0 (1 - \lambda),$$

where λ is a small detuning parameter. We then rescale time by

$$\tau = \nu t,$$

and the equations of motion become

$$\omega_0 \mathbf{x}' = A_0\mathbf{x} + JD_{\mathbf{x}}H_1(\mathbf{x}) + (\delta A_1 + \lambda A_0 + h \cos \tau P)\,\mathbf{x},$$

where the prime denotes differentiation with respect to τ and the parameters δ, λ and $h > 0$ are all small. They are considered as being of the same order of smallness, and higher order terms in the small parameters are neglected.

Since the forcing terms are also Hamiltonian we shall follow the algorithm for obtaining the normal form for a non-autonomous Hamiltonian system given in [17] Very similarly to the unforced case, we find that the quadratic part of the normal form of the unperturbed Hamiltonian function (when $\delta = \lambda = h = 0$) is given by

$$K_0^0 = \frac{1}{\omega_0}\left[\frac{i}{4}(z_1 - \bar{z}_1)^2 - i\Lambda_1 z_2 \bar{z}_2\right].$$

We split this Hamiltonian, and its adjoint (the Hamiltonian associated with A^*)

$$\left(K_0^0\right)^* = \frac{1}{\omega_0}\left[\frac{i}{4}(z_1 + \bar{z}_1)^2 + i\Lambda_1 z_2 \bar{z}_2\right]$$

into semisimple and nilpotent parts,

$$K_S = -\frac{i\Lambda_1}{\omega_0} z_2\bar{z}_2, \qquad K_N = \frac{i}{4\omega_0}\left(z_1 - \bar{z}_1\right)^2,$$

$$K_S^* = \frac{i\Lambda_1}{\omega_0} z_2 z_2, \qquad K_N^* - \frac{i}{4\omega_0}\left(z_1 + \bar{z}_1\right)^2$$

The normal form must belong to the intersection of the kernel of the operators $\mathcal{S}^*$ and $\mathcal{N}^*$, given by

$$\mathcal{S}^* = \frac{i\Lambda_1}{\omega_0}\left(\bar{z}_2\frac{\partial}{\partial\bar{z}_2} - z_2\frac{\partial}{\partial z_2}\right) + \frac{\partial}{\partial\tau}, \qquad \mathcal{N}^* = \frac{i}{2\omega_0}\left(z_1 + \bar{z}_1\right)\left(\frac{\partial}{\partial\bar{z}_1} - \frac{\partial}{\partial z_1}\right)$$

For terms of the form $z_1^{m_1} z_2^{m_2} \bar{z}_1^{m_3} \bar{z}_2^{m_4} e^{is\tau}$ the condition to be in the kernel of $\mathcal{S}^*$ is $s = \Lambda_1\left(m_2 - m_4\right)$. We then calculate the action of the operator $\mathcal{N}^*$ on terms in $\ker\mathcal{S}^*$ (see [17] for the details of this calculation). Then, by calculating the matrix representation of the action of $\mathcal{N}^*$ on $\ker\mathcal{S}^*$, and finding the kernel of this matrix, we can determine the normal form. The result is that the Hamiltonian normal form is given by

$$\begin{aligned} K(\mathbf{z}) \;=\;& \frac{i}{4\omega_0}\left(z_1 - \bar{z}_1\right)^2 + \frac{i\beta_1\delta}{4\omega_0}\left(z_1 + \bar{z}_1\right)^2 - \frac{i}{\omega_0}\left[\Lambda_1\left(1+\lambda\right) - \beta_2\delta\right] z_2\bar{z}_2 \\ &+ \frac{i\alpha_1}{\omega_0}\left(z_1 + \bar{z}_1\right)^4 + \frac{i\alpha_2}{\omega_0}\left(z_1 + \bar{z}_1\right)^2 z_2\bar{z}_2 + \frac{i\alpha_3}{\omega_0} z_2^2\bar{z}_2^2 \\ &+ \underbrace{\frac{h\sigma_1}{\omega_0}(z_1+\bar{z}_1)z_2e^{i\tau} - \frac{h\bar{\sigma}_1}{\omega_0}(z_1+\bar{z}_1)\bar{z}_2e^{-i\tau}}_{\text{combination resonance}} - \underbrace{\frac{h\sigma_4}{2\omega_0}\left(z_2^2e^{i\tau} - \bar{z}_2^2e^{-i\tau}\right)}_{\text{subharmonic resonance}}, \end{aligned} \tag{2.15}$$

where $\sigma_1 = \sigma_1^{re} + i\sigma_1^{im}$ and $\sigma_4 = \sigma_4^{re} + i\sigma_4^{im}$ are complex normal form coefficients. For the shaft (see the Appendix), we have

$$\sigma_1^{re} < 0, \qquad \sigma_1^{im} < 0, \qquad \sigma_4^{re} > 0, \qquad \sigma_4^{im} = 0.$$

The corresponding equations of motion are

$$\begin{aligned} \omega_0 z_1' &= -\frac{i}{2}(z_1 - \bar{z}_1) + \frac{i\beta_1\delta}{2}(z_1 + \bar{z}_1) + 4i\alpha_1(z_1+\bar{z}_1)^3 \\ &\quad + 2i\alpha_2(z_1+\bar{z}_1)z_2\bar{z}_2 + h\sigma_1 z_2 e^{i\tau} - h\bar{\sigma}_1\bar{z}_2e^{-i\tau}, \\ \omega_0 z_2' &= -i[\,\Lambda_1(1+\lambda) - \beta_2\delta\,]z_2 + i\alpha_2(z_1+\bar{z}_1)^2 z_2 + 2i\alpha_3 z_2^2\bar{z}_2 \\ &\quad - h\bar{\sigma}_1(z_1+\bar{z}_1)e^{-i\tau} + h\sigma_4\bar{z}_2e^{-i\tau}. \end{aligned} \tag{2.16}$$

There are two critical cases where the forcing terms occur. If $\omega_0 = \Lambda_1$, there is a combination resonance, and only the forcing terms with coefficients σ_1 are present. On the other hand, if $\omega_0 = 2\Lambda_1$, there is a subharmonic resonance, and only the forcing terms with coefficients σ_4 are present. When the system is near neither of these resonant frequencies, none of the forcing terms are present. We thus have separate equations for the two different cases.

We can further simplify the equations (2.16) by making a time-periodic symplectic coordinate transformation which eliminates nonautonomous terms at leading order. This transformation is given by

$$z_1 = w_1 \qquad z_2 = w_2 e^{-i\tau\omega_0/\Lambda_1} \tag{2.17}$$

The transformation (2.17) also removes the $\mathcal{O}(1)$ linear term from the second modal equation. Finally, we rescale time as $t = \omega_0 \tau$ to eliminate the factor of Λ_1 or $2\Lambda_1$ in each of the equations. We obtain

$$\begin{aligned}\dot{w}_1 &= -\frac{i}{2}(w_1 - \bar{w}_1) + \frac{i\beta_1\delta}{2}(w_1 + \bar{w}_1) + 4i\alpha_1(w_1 + \bar{w}_1)^3 \\ &\quad + 2i\alpha_2(w_1 + \bar{w}_1)w_2\bar{w}_2 + h\sigma_1 w_2 - h\bar{\sigma}_1\bar{w}_2, \\ \dot{w}_2 &= i(-\Lambda_1\lambda + \beta_2\delta)w_2 + i\alpha_2(w_1 + \bar{w}_1)^2 w_2 + 2i\alpha_3 w_2^2\bar{w}_2 \\ &\quad - h\bar{\sigma}_1(w_1 + \bar{w}_1) + h\sigma_4\bar{w}_2.\end{aligned} \tag{2.18}$$

The equations for the two cases (subharmonic and combination resonances) are identical, except for the forcing terms. We point out that time-dependent terms have not been removed, but rather transformed so that time-dependence appears only in higher-order terms. Higher-order terms have been neglected in (2.18). We could also view (2.18) as the result of an averaging procedure.

The nearly symmetric system. For the shaft with a nearly circular cross-section, we have calculated a similar normal form that includes the effects of forcing. These equations are given by

$$\begin{aligned}\omega_0 z_1' &= -i\delta z_1 + i\beta\bar{z}_1 + iA_0 z_1^2\bar{z}_1 + iB_0 z_1 z_2\bar{z}_2 - \frac{h}{2}\bar{\sigma}_1\bar{z}_2 e^{-i\tau}, \\ \omega_0 z_2' &= i\left[2\bar{\omega}(1+\lambda) + \delta - \frac{\beta^2}{4\bar{\omega}}\right] z_2 + iB_0 z_1\bar{z}_1 z_2 + iC_0 z_2^2\bar{z}_2 \\ &\quad - \beta h\bar{\sigma}_4\bar{z}_2 e^{-i\tau} - \frac{h}{2}\bar{\sigma}_1\bar{z}_1 e^{-i\tau}.\end{aligned} \tag{2.19}$$

We have added forcing terms with complex coefficients σ_1 and σ_4. As in the general case, the forcing terms with coefficients σ_4 only appear for the subharmonic case, while the forcing terms with coefficients σ_1 only appear for the combination resonance case. We have also added the detuning parameter λ in the same way as for the general case. See McDonald [16] for details on the normal form calculations for the forcing terms.

2.4 Normal forms for the damped system. We now modify the equations of motion by adding terms that model damping. As in the previous section, we begin by considering the unforced system. The original equation of motion (2.1) becomes

$$\ddot{\mathbf{q}} + 2G\dot{\mathbf{q}} + K\mathbf{q} + \nabla_{\mathbf{q}}U(\mathbf{q}) + \zeta[\,\Lambda\dot{\mathbf{q}} + F(\mathbf{q}, \dot{\mathbf{q}})\,] = 0, \tag{2.20}$$

where the 2×2 matrix Λ and the function F represent the linear and nonlinear effects of energy dissipation due to damping, and ζ is a small nonnegative parameter. The matrix Λ is symmetric and positive definite, and the nonlinear function F is linear in its second argument.

Writing the equations of motion in terms of $\mathbf{x} = (\mathbf{q}, \mathbf{p}) \in \mathbb{R}^4$ we obtain

$$\dot{\mathbf{x}} = A\mathbf{x} + JD_{\mathbf{x}}H_1(\mathbf{x}) - \zeta[\,B\mathbf{x} + f(\mathbf{x})\,], \tag{2.21}$$

where

$$B = \begin{bmatrix} O_2 & O_2 \\ -\Lambda G & \Lambda \end{bmatrix}$$

and

$$f(\mathbf{x}) = \begin{bmatrix} \mathbf{0}_2 \\ F(\mathbf{q}, \mathbf{p} - G\mathbf{q}) \end{bmatrix}.$$

Here O_2 is the 2×2 null matrix and $\mathbf{0}_2$ is the null vector in $\mathbb{R}^2$. The equation of motion (2.21) is no longer Hamiltonian when $\zeta > 0$, due to the dissipative terms. As explained in the Appendix, the damping terms in the equations of motion of gyroscopic systems, in general, can be due to internal damping (Kelvin-Voigt visco-elastic material) and external damping (friction between the structure and its supports).

The first step is to apply the transformation $\mathbf{y} = \hat{T}\mathbf{x}$, where $\hat{T} = T\left(I + \tilde{T}\right)$, that normalizes the linear Hamiltonian terms in the equations of motion, including the damping terms. Then, we calculate the normal form for linear terms in the equation of motion. For our purposes, we also need to calculate the matrix T_D i.e. the matrix which transforms the linear damping terms into their normal form. This transformation affects the nonlinear Hamiltonian terms, creating terms which are at the same order as the nonlinear damping terms. Thus, we must calculate the matrix T_D and calculate the effect of applying the transformation $I + \zeta^* T_D$ on the nonlinear Hamiltonian terms. The normal form for the linear damping terms can be written as

$$\delta_1 \begin{bmatrix} z_1 \\ 0 \\ \bar{z}_1 \\ 0 \end{bmatrix} + \delta_2 \begin{bmatrix} 0 \\ z_2 \\ 0 \\ 0 \end{bmatrix} + \delta_3 \begin{bmatrix} z_1 + \bar{z}_1 \\ 0 \\ -(z_1 + \bar{z}_1) \\ 0 \end{bmatrix} + \delta_4 \begin{bmatrix} 0 \\ 0 \\ 0 \\ \bar{z}_2 \end{bmatrix}$$

Using reality conditions, we can show that $\delta_4 = \bar{\delta}_2$, δ_1 is real, and δ_3 is imaginary. For the shaft, we find (see the Appendix) that δ_2 is real, and

$$\delta_1 < 0, \qquad \delta_2 < 0, \qquad \delta_3 = 0.$$

Normal form coefficients for nonlinear terms (cubic order in $\mathbf{z}$) can also be calculated. See [18] for the form of the nonlinear damping terms.

The equations of motion for the system with damping become

$$\begin{aligned} \dot{z}_1 &= -\frac{i}{2}(z_1 - \bar{z}_1) + \frac{i\beta_1\delta}{2}(z_1 + \bar{z}_1) + \zeta\delta_1 z_1 + \zeta\delta_3(z_1 + \bar{z}_1) \\ &\quad + 4i\alpha_1(z_1 + \bar{z}_1)^3 + 2i\alpha_2(z_1 + \bar{z}_1)z_2\bar{z}_2, \\ \dot{z}_2 &= i(-\Lambda_1 + \beta_2\delta)z_2 + \zeta\delta_2 z_2 + i\alpha_2(z_1 + \bar{z}_1)^2 z_2 + 2i\alpha_3 z_2^2\bar{z}_2. \end{aligned} \tag{2.22}$$

in the unforced case, and

$$\begin{aligned} \dot{w}_1 &= -\frac{i}{2}(w_1 - \bar{w}_1) + \frac{i\beta_1\delta}{2}(w_1 + \bar{w}_1) + \zeta\delta_1 w_1 + \zeta\delta_3(w_1 + \bar{w}_1) \\ &\quad + 4i\alpha_1(w_1 + \bar{w}_1)^3 + 2i\alpha_2(w_1 + \bar{w}_1)w_2\bar{w}_2 + h\sigma_1 w_2 - h\bar{\sigma}_1\bar{w}_2, \\ \dot{w}_2 &= i(-\Lambda_1\lambda + \beta_2\delta)w_2 + \zeta\delta_2 w_2 + i\alpha_2(w_1 + \bar{w}_1)^2 w_2 + 2i\alpha_3 w_2^2\bar{w}_2 \\ &\quad - h\bar{\sigma}_1(w_1 + \bar{w}_1) + h\sigma_4\bar{w}_2. \end{aligned} \tag{2.23}$$

in the forced case, the latter after rescaling time and using the periodic coordinate transformation (2.17).

The nearly symmetric system. For the shaft with a nearly circular cross-section, normal form calculations are simplified due to the approximate S^1-symmetry. If the damping parameter ζ, the symmetry-breaking parameter β and the bifurcation parameter δ are all treated as being of the same order of smallness (i.e. ζ, $|\delta|$, $\beta \ll$

1), or if we take $\zeta \ll |\delta|$, $\beta \ll 1$, it suffices to take the leading order of approximation

$$\Lambda = \begin{bmatrix} \lambda_0 & 0 \\ 0 & \lambda_0 \end{bmatrix}$$

for the linear damping term in (2.21), where $\lambda_0 > 0$. This linear damping term is S^1-symmetric, and similarly it suffices to take the nonlinear damping term $f(\mathbf{x})$ as S^1-symmetric.

To transform (2.21) into normal form, we begin by using the same (symplectic) change of basis matrix $T = T(\delta, \tilde{\beta})$ used in the undamped case (2.11) to calculate the matrix for the linear part of the transformed equation with damping,

$$T^{-1}[A - \zeta B]T.$$

Then a number of successive coordinate changes simplify the linear and nonlinear terms (details in [22]), yielding the normal form for the unforced damped nearly symmetric system:

$$\begin{aligned} \dot{z}_1 &= \delta\left(-i + \frac{\zeta\lambda_0}{2\bar{\omega}}\right) z_1 + \beta\left(i - \frac{\zeta\lambda_0}{2\bar{\omega}}\right) \bar{z}_1 \\ &\quad + (iA_0 + \zeta\alpha_{11}) z_1^2 \bar{z}_1 + (iB_0 + \zeta\alpha_{12}) z_1 z_2 \bar{z}_2, \\ \dot{z}_2 &= -\zeta\lambda_0 z_2 + i\left(2\bar{\omega} + \delta - \frac{\beta^2}{4\bar{\omega}}\right) z_2 \\ &\quad + (iB_0 + \zeta\alpha_{21}) z_1 \bar{z}_1 z_2 + (iC_0 + \zeta\alpha_{22}) z_2^2 \bar{z}_2, \end{aligned} \tag{2.24}$$

where the normal form coefficients α_{jk} are real. These latter coefficients depend on λ_0, the nonlinear damping function $F(\mathbf{q}, \dot{\mathbf{q}})$ and the nonlinear conservative term $U(\mathbf{q})$. The values of these nonlinear coefficients can be found in Nagata and Namachchivaya [22] for the rotating shaft problem. The normal form for the forced damped nearly symmetric system is similar, except for the addition of forcing terms as in (2.19).

3 The unforced undamped system

In this section we consider the normal form for the unforced undamped system and study the dynamics, both for the general and nearly symmetric cases. It is convenient to express the first mode in rectangular coordinates and the second mode in action-angle variables:

$$z_1 = x_1 + iy_1, \qquad z_2 = \sqrt{2I_2}\, e^{i\theta_2}. \tag{3.1}$$

In these coordinates, the equations of motion (2.9) become

$$\begin{aligned} \dot{x}_1 &= \beta_1 \delta y_1 + 32\alpha_1 y_1^3 + 8\alpha_2 y_1 I_2, \\ \dot{y}_1 &= x_1, \\ \dot{I}_2 &= 0, \\ \dot{\theta}_2 &= -\Lambda_1 + \beta_2\delta + 4\alpha_3 I_2, \end{aligned} \tag{3.2}$$

We have nontrivial equilibrium solutions if

$$I_2 = 0, \qquad x_1 = 0, \qquad \beta_1\delta + 32\alpha_1 y^2 = 0, \tag{3.3}$$

which bifurcate from the origin at $\delta = 0$. For the shaft, $\beta_1 > 0$ and $\alpha_1 < 0$ so the bifurcation is supercritical. We call these *first mode solutions* $PS1$ since they exist in the x_1y_1-plane, $I_2 = 0$. The $PS1$ can be shown to be Lyapunov stable.

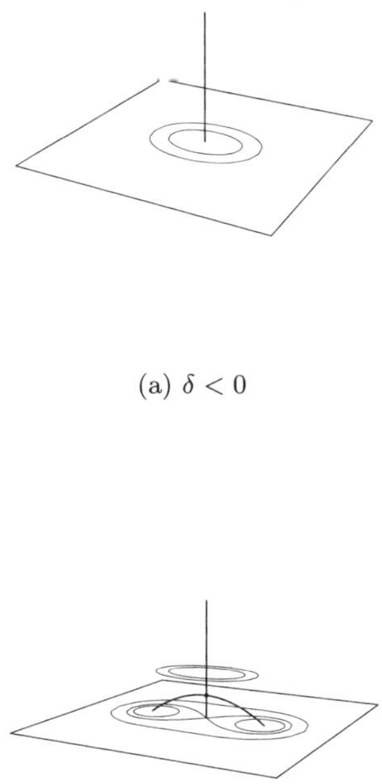

(a) $\delta < 0$

(b) $\delta > 0$

Figure 2 Phase portraits for the unforced undamped system (3.2).

The global dynamics of the unforced undamped system are easily obtained. From the third equation in (3.2), we see that I_2 is constant, and therefore can be treated as a parameter in the first two equations. The first two equations with constant I_2 comprise a one degree-of-freedom Hamiltonian system, with Hamiltonian function

$$\frac{1}{2}x_1^2 - \left(\frac{1}{2}\beta_1\delta + 4\alpha_2 I_2\right)y_1^2 - 8\alpha_1 y_1^4.$$

By examining the level sets of this Hamiltonian function, we easily obtain the phase portraits shown in Figure 2 for the shaft.

Note that in the three-dimensional representations of Figure 2, what appear as equilibria along the positive I_2-axis (shown protruding vertically from the horizontal x_1y_1-plane) are actually periodic orbits in the four-dimensional phase space of (3.2) with angular velocity $\dot{\theta}_2 \approx -\Lambda_1$ near the origin. This is more apparent in rectangular coordinates for the second mode, $z_2 = x_2 + iy_2$.

For $\delta < 0$ there is a unique stable equilibrium at the origin, two families of periodic orbits (one family in the x_1y_1-plane, the other family represented by the positive I_2-axis), and the rest of the orbits lie on two-tori in the four-dimensional

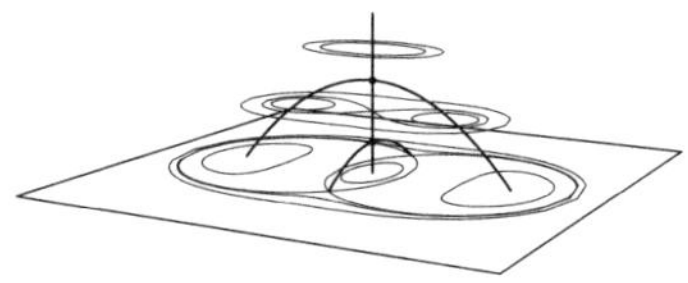

Figure 3 Phase portrait for the nearly symmetric unforced undamped system (3.4) with $\delta > \beta$.

phase space. For $\delta > 0$ the origin has become an unstable equilibrium, and the pair of stable $PS1$ solutions lie in the x_1y_1-plane. In the x_1y_1 plane there is a pair of orbits homoclinic to the origin. For $0 < I_2 < \beta_1\delta/(8|\alpha_2|)$ there are also pairs of homoclinic orbits, but these are homoclinic to periodic orbits in the four-dimensional phase space. Besides the equilibria and homoclinic orbits, there are families of periodic orbits and two-tori.

The nearly symmetric system. In the same way as described just above, we obtain the equilibria and phase portraits for the unforced undamped system in the nearly symmetric case. In the $(x_1, y_1, I_2, \theta_2)$ coordinates, (2.13) becomes

$$\begin{aligned}
\dot{x}_1 &= (\delta+\beta)y_1 - A_0(x_1^2+y_1^2)y_1 - 2B_0I_2y_1,\\
\dot{y}_1 &= -(\delta-\beta)x_1 + A_0(x_1^2+y_1^2)x_1 + 2B_0I_2x_1,\\
\dot{I}_2 &= 0,\\
\dot{\theta}_2 &= 2\bar{\omega} + \delta + \frac{\beta^2}{4\bar{\omega}} + B_0(x_1^2+y_1^2) + 2C_0I_2,
\end{aligned} \tag{3.4}$$

For the nearly symmetric shaft, the phase portrait of (3.4) for $\delta < -\beta$ is the same as Figure 2 (a), and the phase portrait of (3.4) for $-\beta < \delta < \beta$ is the same as Figure 2 (b): the stable first mode solutions $PS1$ bifurcate supercritically from the origin at $\delta = -\beta$. For $\delta > \beta$ there is an additional phase portrait for (3.4), shown in Figure 3.

As δ increases past the value of β, the first mode solutions $PS1$ become unstable saddles, and a new pair of stable first mode solutions $PS1'$ bifurcate from the origin, and the origin itself regains Lyapunov stability.

4 The unforced damped system

In this section, we consider the effects of damping in the unforced system for both the general and nearly symmetric cases. In the coordinates $(x_1, y_1, I_2, \theta_2)$

given by (3.1), the equations of motion (2.22) become

$$\begin{aligned}\dot{x}_1 &= \beta_1\delta y_1 + 32\alpha_1 y_1^3 + 8\alpha_2 y_1 I_2 + \zeta\delta_1 x_1,\\ \dot{y}_1 &= x_1 + \zeta\delta_1 y_1,\\ \dot{I}_2 &= 2\zeta\delta_2 I_2,\\ \dot{\theta}_2 &= -\Lambda_1 + \beta_2\delta + 4\alpha_3 I_2.\end{aligned} \tag{4.1}$$

From the third equation in (4.1) we see that the x_1y_1-plane $I_2 = 0$ is invariant and exponentially attracting (since $\zeta > 0$ and $\delta_2 < 0$). Thus there are no equilibrium solutions with $I_2 > 0$ (i.e. no second mode or mixed-mode solutions). On the invariant plane of the first mode, the dynamics are described by the first two equations of (4.1) with $I_2 = 0$.

The trivial solution becomes unstable at $\delta = \zeta^2\delta_1^2/\beta_1$. At this point, we have a bifurcation from the trivial solution, of first mode solutions which we again denote by $PS1$. The solutions $PS1$ are given by

$$I_2 = 0, \qquad x_1 = -\zeta\delta_1 y_1, \qquad \beta_1\delta - \zeta^2\delta_1^2 + 32\alpha_1 y^2 = 0. \tag{4.2}$$

Comparing with the undamped case (3.3), we see that for $\zeta \ll 1$ there is little change. For the shaft, the bifurcation is supercritical and the $PS1$ solutions are stable. In the damped case, the stable solutions are asymptotically stable as well as Liapunov stable.

If nonlinear damping terms are added, the local dynamics remain essentially the same. There are no second mode or mixed-mode solutions, and there is a supercritical bifurcation of first mode solutions $PS1$ from the trivial solution at $\delta = \zeta^2\delta_1^2/\beta_1$. The $PS1$ solutions are stable near the bifurcation point. At large amplitudes there may be secondary Hopf bifurcations from the $PS1$, see [18]. But since these are at large amplitude ($|(x_1, y_1)| = \mathcal{O}(1)$), they may not persist if all higher order terms neglected in (4.1) are restored.

The nearly symmetric system. For the nearly symmetric system, the normal form (2.24) written in the $(x_1, y_1, I_2, \theta_2)$ coordinates given by (3.1) becomes

$$\begin{aligned}\dot{x}_1 &= \frac{\zeta\lambda_0}{2\bar{\omega}}(\delta-\beta)x_1 + (\delta+\beta)y_1 + (x_1^2+y_1^2)(\zeta\alpha_{11}x_1 - A_0y_1) + 2I_2(\zeta\alpha_{12}x_1 - B_0y_1),\\ \dot{y}_1 &= -(\delta-\beta)x_1 + \frac{\zeta\lambda_0}{2\bar{\omega}}(\delta+\beta)y_1 + (x_1^2+y_1^2)(A_0x_1 + \zeta\alpha_{11}y_1) + 2I_2(B_0x_1 + \zeta\alpha_{12}y_1),\\ \dot{I}_2 &= -\zeta[\,\lambda_0 - \alpha_{21}(x_1^2+y_1^2) - 2\alpha_{22}I_2\,]\,I_2,\\ \dot{\theta}_2 &= 2\bar{\omega} + \delta + \frac{\beta^2}{4\bar{\omega}} + B_0(x_1^2+y_1^2) + 2C_0I_2,\end{aligned} \tag{4.3}$$

Again we observe that the x_1y_1-plane $\{I_2 = 0\}$ is invariant and exponentially attracting, at least in a neighbourhood of the origin (since $\zeta > 0$, $\lambda_0 > 0$). Thus the dynamics reduce to the planar system consisting of the first two equations of (4.3) with $I_2 = 0$. One can then find the equilibrium solutions and determine their linearized stabilities, and analyze local bifurcations. Details are given in [22], but here we mention two items of interest. The first is that there is dissipation-induced instability. If damping is absent, then the trivial equilibrium is stable for the Hamiltonian system when $\delta > \beta$, but when damping is present the same equilibrium is unstable. The other item of interest is that the small parameter ζ cannot be treated as being of the same order of smallness as the other small parameters δ and

Figure 4 Phase portraits for the nearly symmetric unforced damped system.

β. In (2.24), Hamiltonian terms of order $\mathcal{O}(|\delta,\beta|^2|z| + |\delta,\beta||z|^3 + |z|^5)$ have been neglected, but dissipative terms of order $\mathcal{O}(|\zeta||\delta,\beta||z|+|\zeta||z|^3)$ have been kept. The latter terms are of the lowest order that give structurally stable bifurcations; if they are neglected then the x_1y_1-plane is still exponentially attracting, but the reduced dynamics on the x_1y_1-plane become Hamiltonian, and are therefore non-generic and structurally unstable.

Since the reduced system is planar, and Hamiltonian if $\zeta = 0$, one can use Melnikov's method to determine the global dynamics for small $\zeta > 0$. For $A_0 > 0$, we obtain the phase portraits in Figure 4, but for the actual shaft parameter values, only phase portraits A, B and J are realized [22].

5 The forced undamped system – subharmonic resonance case

In this section, we consider the forced system with no damping, the normal form (2.18). We consider only the subharmonic case, with $\sigma_1 = 0$. In this case, the forcing only enters into the equations of motion in the second mode. Because the first mode is marginally unstable and the second mode is excited, both modes can bifurcate from the trivial solution, and interact with each other.

In rectangular coordinates

$$w_1 = x_1 + iy_1, \qquad w_2 = x_2 + iy_2, \tag{5.1}$$

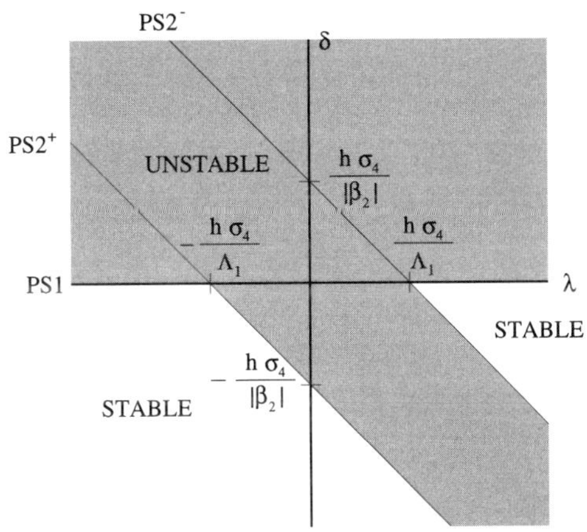

(a) shaft

Figure 5 Stability boundaries for the trivial solution for the forced undamped shaft.

the equations of motion (2.18) are given by

$$\begin{aligned}
\dot{x}_1 &= \beta_1 \delta y_1 + 32\alpha_1 y_1^3 + 4\alpha_2 y_1 \left(x_2^2 + y_2^2\right), \\
\dot{y}_1 &= x_1, \\
\dot{x}_2 &= \left(\beta_2 \delta - \lambda\Lambda_1\right) y_2 + 4\alpha_2 y_1^2 y_2 + 2\alpha_3 y_2 \left(x_2^2 + y_2^2\right) - h\sigma_4^{re} x_2 + h\sigma_4^{im} y_2, \\
\dot{y}_2 &= -\left(\beta_2 \delta - \lambda\Lambda_1\right) x_2 - 4\alpha_2 y_1^2 x_2 - 2\alpha_3 x_2 \left(x_2^2 + y_2^2\right) + h\sigma_4^{re} y_2 + h\sigma_4^{im} x_2.
\end{aligned} \tag{5.2}$$

Note that for the shaft we have $\sigma_4^{im} = 0$.

The eigenvalues of the linearization of (5.2) about the trivial solution are

$$\lambda_{1,2} = \pm\sqrt{\beta_1 \delta}, \qquad \lambda_{3,4} = \pm i\sqrt{\left(\beta_2 \delta - \lambda\Lambda_1\right)^2 - h^2\sigma_4^2}$$

For simplicity, we have defined

$$\sigma_4^2 = \left(\sigma_4^{re}\right)^2$$

where $\sigma_4 > 0$. The first pair of eigenvalues is imaginary for $\delta < 0$, zero at $\delta = 0$, and real for $\delta > 0$. Thus, the system is unstable for $\delta > 0$. Since $\beta_2 < 0$ for the shaft, we have

$$\text{eigenvalues imaginary}: \qquad \delta < \frac{-\lambda\Lambda_1 - h\sigma_4}{|\beta_2|} \text{ or } \delta > \frac{-\lambda\Lambda_1 + h\sigma_4}{|\beta_2|}$$

$$\text{eigenvalues real}: \qquad \frac{-\lambda\Lambda_1 - h\sigma_4}{|\beta_2|} < \delta < \frac{-\lambda\Lambda_1 + h\sigma_4}{|\beta_2|}$$

Thus, we can determine the stability of the trivial solution as a function of δ and λ. This behavior is shown in Figure 5, where the unshaded regions are stable, and the shaded regions are unstable. For a Hamiltonian system such as this, the eigenvalues are symmetric about the real and imaginary axes. Thus, the trivial solution is stable when all eigenvalues are on the imaginary axis, and unstable when at least one pair of eigenvalues is real.

We see an interesting phenomenon here. In gyroscopic systems, such as the one studied by Nagata and Namachchivaya [22], the pair of eigenvalues corresponding to the first mode may re-stabilize after initially becoming unstable. Since we are only looking at the normal form near the first critical point of the first mode, we do not see the re-stabilization locally in the eigenvalues$\lambda_{1,2}$. However, for the second pair of eigenvalues, $\lambda_{3,4}$ we see an instability and re-stabilization for the case $\lambda > \frac{\sigma_4}{\Lambda_1}$. Without the forcing, the two eigenvalues $\lambda_{3,4}$ would not destabilize, and since there was no time-dependent transformation, they would be located near $\pm i\Lambda_1$. With forcing, and the associated time-dependent transformation to eliminate the explicit time-dependence of the forcing terms, these eigenvalues are located near zero, and undergo an instability and re-stabilization similar to the instability that the first pair of eigenvalues, $\lambda_{1,2}$ may undergo. The loss of stability in this mode is due to the symmetry-breaking caused by the subharmonic resonant forcing. The forcing breaks the S^1 symmetry of the second mode, causing instability, in much the same way as the symmetry-breaking in the first mode causes instability in that mode.

In the remainder of this section, we describe the local bifurcation and stability of equilibrium solutions of (2.18). Details are similar to those given in the pipes application [18], so are not given here.

First mode solutions. We first look for single-mode solutions to the nonlinear equations. We find a pair of first mode equilibrium solutions, which we denote by $PS1$, by setting $x_2 = y_2 = 0$ in (5.2) and then solving $\dot{x}_1 = 0$, $\dot{y}_1 = 0$. We then determine the linearized stability of the $PS1$ solutions.

We find that the $PS1$ solutions only exist for $\delta > 0$ and are supercritical. We further determine that $PS1$ may have 0, 1, or 2 bifurcation points, and for a fixed value of λ, we have three cases:

1. For $\lambda > h\sigma_4/\Lambda_1$, no bifurcation points exist on $PS1$, and the solution $PS1$ is always stable.
2. For $-h\sigma_4/\Lambda_1 < \lambda < h\sigma_4/\Lambda_1$, only one bifurcation point B_2 exists, and the solution is unstable for $\delta < \delta_{B_2}$, and stable for $\delta > \delta_{B_2}$.
3. For $\lambda < -h\sigma_4/\Lambda_1$, two bifurcation points B_1 and B_2 exist, and the solution is unstable between the two bifurcations, and stable outside the two bifurcations.

Second mode solutions. By setting $x_1 = y_1 = 0$, we find two distinct pairs of second mode solutions for (5.2), the pairs denoted $PS2^-$ and $PS2^+$. The $PS2^+$ pair of solutions bifurcate from the trivial solution into the x_2y_2-plane at

$$\delta = \frac{\lambda\Lambda_1 + h\sigma_4}{\beta_2},$$

while $PS2^-$ bifurcate from the trivial solution at

$$\delta = \frac{\lambda\Lambda_1 - h\sigma_4}{\beta_2},$$

For the shaft, both bifurcations of $PS2$ solutions are subcritical. Furthermore, we have $\beta_2 < 0$, thus the bifurcation to $PS2^+$ occurs first, as δ is increased.

A critical question to consider is whether the first mode solution or the second mode solutions bifurcate from the trivial solution first. The order in which these bifurcations occurs as δ is increased is summarized in Table 1. Figure 5 shows not only the stability of the trivial solution, but where each of the bifurcations to modes $PS1$, $PS2^-$, and $PS2^+$ may occur for the shaft.

Next, we can determine the stability of the second mode solutions. We find:

Table 1 Order of bifurcations of trivial solution as the parameter δ is increased.

parameter region	$(\beta_2 > 0)$	$(\beta_2 < 0)$
$\lambda < -h\sigma_4/\Lambda_1$	$\{PS2^-, PS2^+, PS1\}$	$\{PS1, PS2^+, PS2^-\}$
$-h\sigma_4/\Lambda_1 < \lambda < h\sigma_4/\Lambda_1$	$\{PS2^-, PS1, PS2^+\}$	$\{PS2^+, PS1, PS2^-\}$
$\lambda > h\sigma_4/\Lambda_1$	$\{PS1, PS2^-, PS2^+\}$	$\{PS2^+, PS2^-, PS1\}$

1. For $\lambda > h\sigma_4/\Lambda_1$, $PS2^\pm$ both bifurcate after $PS1$. Neither of $PS2^\pm$ has bifurcations, $PS2^+$ is always unstable, and $PS2^-$ is always stable.
2. For $-h\sigma_4/\Lambda_1 < \lambda < h\sigma_4/\Lambda_1$, $PS2^+$ bifurcates before $PS1$, and $PS2^-$ bifurcates after $PS1$. $PS2^-$ has a bifurcation at B_3^-, and is stable before that bifurcation, and unstable afterwards. $PS2^+$ has no bifurcations, and is always unstable.
3. For $\lambda < -h\sigma_4/\Lambda_1$, $PS1$ bifurcates before both $PS2^\pm$. Solution $PS2^+$ has a bifurcation point B_3^+ and is always unstable. $PS2^-$ has a bifurcation point B_3^-, and is stable before the bifurcation, and unstable afterwards.

These scenarios are shown in Figure 6. The second mode solutions appear as equilibrium solutions of (5.2), but in the original normal form (2.16) they correspond to small-amplitude subharmonic time-periodic solutions of period π, that have been excited by the forcing terms when $h > 0$.

Multi-mode solutions. Finally, we can determine the multi-mode solutions $MM1$ and $MM2$ for this case, i.e. solutions of (5.2) for which the amplitude of both modes are non-zero. For the shaft, we can also determine that solution $MM1$ connects B_2, the second bifurcation point of $PS1$ to B_3^-, the bifurcation point of $PS2^-$, and $MM2$ connects B_1, the first bifurcation point of $PS1$ to B_3^+, the bifurcation point of $PS2^+$. The existence of each multi-mode solution has the same criteria as the corresponding existence criteria of the bifurcation points that are its end points. Namely, for $\lambda > h\sigma_4/\Lambda_1$, no multi-mode solutions exist, for $-h\sigma_4/\Lambda_1 < \lambda < h\sigma_4/\Lambda_1$, only $MM1$ exists, and for $\lambda < -h\sigma_4/\Lambda_1$, both $MM1$ and $MM2$ exist. The multi-mode solutions are shown in Figure 6.

The stability of these multi-mode solutions is difficult to determine analytically, but we can easily determine it numerically for specific cases. We also note that the multi-mode solutions provide a mechanism for energy to be transferred from the high frequency second mode to the low frequency first mode. For values of δ at which $MM1$ exists, the $MM1$ solutions are stable, and the branch of $MM1$ solutions connects the branches of stable $PS2^-$ solutions and the stable $PS1$ solutions. This fact can be interpreted as an energy transfer between the modes.

6 The forced damped system – local dynamics

In this section, we consider local bifurcations in the system with both forcing and damping, in the subharmonic case. The equations of motion (2.23) with $\sigma_1 = 0$

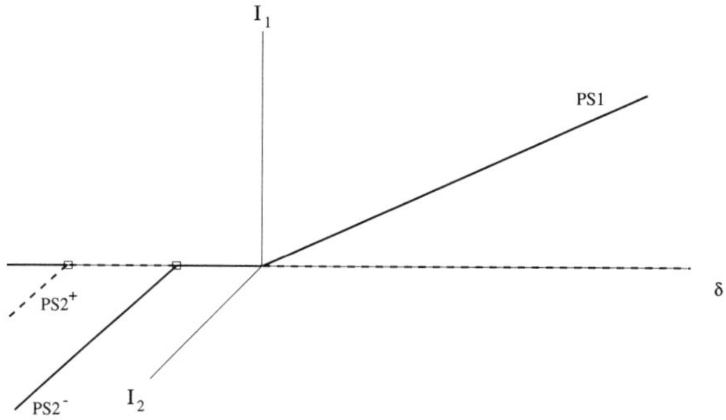

(a) $\lambda > \frac{h\sigma_4}{\Lambda_1}$

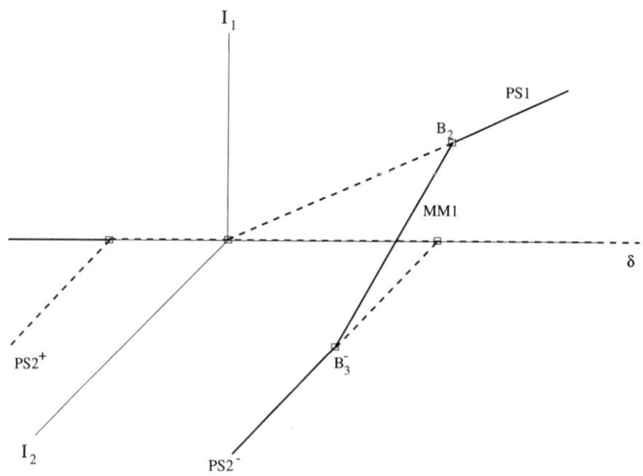

(b) $-\frac{h\sigma_4}{\Lambda_1} < \lambda < \frac{h\sigma_4}{\Lambda_1}$

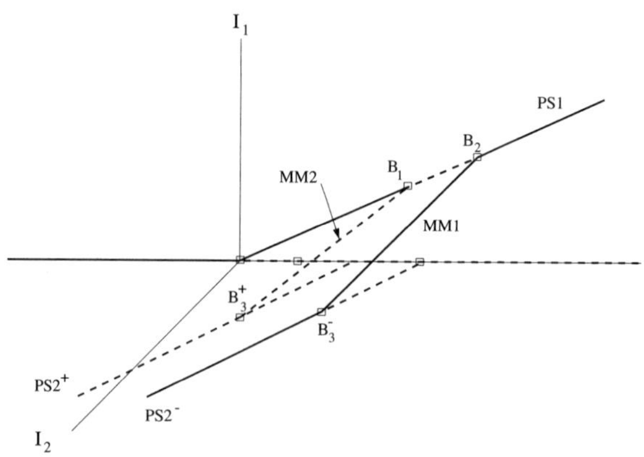

(c) $\lambda < -\frac{h\sigma_4}{\Lambda_1}$

Figure 6 Bifurcation scenarios for the forced undamped rotating shaft.

(the subharmonic case), in rectangular coordinates (5.1) are

$$\begin{aligned}
\dot{x}_1 &= \beta_1\delta y_1 + 32\alpha_1 y_1^3 + 4\alpha_2 y_1\left(x_2^2+y_2^2\right) + \zeta\delta_1 x_1,\\
\dot{y}_1 &= x_1 + \zeta\delta_1 y_1,\\
\dot{x}_2 &= \left(\beta_2\delta - \lambda\Lambda_1\right) y_2 + 4\alpha_2 y_1^2 y_2 + 2\alpha_3 y_2\left(x_2^2+y_2^2\right) + \zeta\delta_2 x_2 - h\sigma_4^{re} x_2 + h\sigma_4^{im} y_2,\\
\dot{y}_2 &= -\left(\beta_2\delta - \lambda\Lambda_1\right) x_2 - 4\alpha_2 y_1^2 x_2 - 2\alpha_3 x_2\left(x_2^2+y_2^2\right) + \zeta\delta_2 y_2 + h\sigma_4^{re} y_2 + h\sigma_4^{im} x_2.
\end{aligned} \tag{6.1}$$

First, we study the stability of the trivial solution. The Jacobian matrix at the origin is given by

$$\begin{bmatrix}
\zeta\delta_1 & \beta_1\delta & 0 & 0\\
1 & \zeta\delta_1 & 0 & 0\\
0 & 0 & \zeta\delta_2 - \sigma_4^{re} & \beta_2\delta - \lambda\Lambda_1 + \sigma_4^{im}\\
0 & 0 & -\beta_2\delta + \lambda\Lambda_1 + \sigma_4^{im} & \zeta\delta_2 + \sigma_4^{re}
\end{bmatrix}$$

By examining the characteristic equations, we can determine that the system undergoes a simple bifurcation to a pair of first mode $PS1$ solutions into the x_1y_1-plane at

$$\delta = \frac{\zeta^2\delta_1^2}{\beta_1}$$

The system also undergoes simple bifurcations of second mode solutions in the x_2y_2-plane at

$$\delta = \frac{\lambda\Lambda_1 + \sqrt{h^2\sigma_4^2 - \zeta^2\delta_2^2}}{\beta_2}$$

for $PS2^+$ solutions, and at

$$\delta = \frac{\lambda\Lambda_1 - \sqrt{h^2\sigma_4^2 - \zeta^2\delta_2^2}}{\beta_2}$$

for $PS2^-$ solutions.

If $\zeta \ll h$, then the bifurcation points differ only slightly from the corresponding bifurcation points for the forced undamped system. We also find in this case that the secondary bifurcation points, corresponding to the bifurcation of multi-mode solutions $MM1$ and $MM2$ similarly exist and differ only slightly from the corresponding points in the forced undamped system. The three bifurcation scenarios depicted in Figure 6 are essentially unchanged when light damping is added, except the bifurcation points move slightly from their locations in the undamped case, and Liapunov stability in the undamped case becomes asymptotic stability in the damped case. Details of the calculations are analogous to those given for pipes in [18].

When nonlinear damping terms are also considered, the local dynamics remain qualitatively the same. Exact analytical results are more difficult to obtain, but bifurcation diagrams can be obtained numerically. As in the unforced damped case, there may be a Hopf bifurcation of the first mode solution. Also, the two second mode solutions may connect. But these effects occur at large amplitudes of the solutions, and may not be stable with respect to the addition of other higher order terms. See [18] for more details in the context of the pipes application.

7 The forced damped system – global dynamics

In this section, we study global dynamics of the normal form (2.23) for the forced damped system. We focus on the subharmonic resonance case ($\sigma_1 = 0$) since the methods we use are not applicable in the combination resonance case. We use the theory developed by Kovačič and Wiggins [11] to study the existence of orbits homoclinic to resonances. Much of the analysis is similar to that in the pipes application [19], so we omit details here.

The equations of motion (2.23) with $\sigma_1 = 0$, in coordinates given by

$$w_1 = x + iy, \qquad w_2 = \sqrt{2I}\, e^{i\theta}, \tag{7.1}$$

with parameter scaling

$$h = \epsilon, \qquad \zeta = \epsilon\hat{\zeta}, \tag{7.2}$$

become

$$\begin{aligned}
\dot{x} &= \beta_1\delta y + 32\alpha_1 y^3 + 8\alpha_2 I y + \epsilon\hat{\zeta}\delta_1 x,\\
\dot{y} &= x + \epsilon\hat{\zeta}\delta_1 y,\\
\dot{I} &= \epsilon[\, 2\hat{\zeta}\delta_2 + 2\sigma_4 \sin 2\phi\,]\, I,\\
\dot{\phi} &= \beta_2\delta - \lambda\Lambda_1 + 4\alpha_2 y^2 + 4\alpha_3 I + \epsilon\sigma_4 \cos 2\phi,
\end{aligned} \tag{7.3}$$

where we have also used

$$\sigma_4^{im} \cos 2\theta - \sigma_4^{re} \sin 2\theta = \sigma_4 \cos 2\phi$$

with

$$\phi = \theta - \frac{D}{2}, \qquad \sigma_4 = [\, (\sigma_4^{re})^2 + (\sigma_4^{im})^2\,]^{1/2}, \qquad \cos D = \frac{\sigma_4^{im}}{\sigma_4}.$$

For the shaft, recall that $\sigma_4^{im} = 0$. The equations (7.3) have the standard form

$$\begin{aligned}
\dot{\mathbf{x}} &= JD_{\mathbf{x}}H_0(\mathbf{x}, I) + \epsilon\left\{JD_{\mathbf{x}}H_1(\mathbf{x}, I, \phi; \epsilon) + f^{\mathbf{x}}(\mathbf{x}, I, \phi; \epsilon)\right\},\\
\dot{I} &= \epsilon\left\{-D_\phi H_1(\mathbf{x}, I, \phi; \epsilon) + f^I(\mathbf{x}, I, \phi; \epsilon)\right\},\\
\dot{\phi} &= D_I H_0(\mathbf{x}, I) + \epsilon\left\{D_I H_1(\mathbf{x}, I, \phi; \epsilon) + f^\phi(\mathbf{x}, I, \phi; \epsilon)\right\},
\end{aligned} \tag{7.4}$$

where $\mathbf{x} = \begin{bmatrix} x & y \end{bmatrix}^T \in \mathbb{R}^2$, $I \in [0, \infty)$, $\phi \in S^1$. The unperturbed Hamiltonian H_0 and perturbed Hamiltonian H_1 for this system are given by

$$\begin{aligned}
H_0 &= -\frac{1}{2}x^2 + \frac{1}{2}\beta_1\delta y^2 + (\beta_2\delta - \lambda\Lambda_1)\, I + 8\alpha_1 y^4 + 4\alpha_2 I y^2 + 2\alpha_3 I^2,\\
H_1 &= \sigma_4 I \cos 2\phi,
\end{aligned}$$

and the dissipative perturbations are given by

$$\begin{aligned}
f^{\mathbf{x}} &= \begin{bmatrix} \hat{\zeta}\delta_1 x \\ \hat{\zeta}\delta_1 y \end{bmatrix},\\
f^I &= 2\hat{\zeta}\delta_2 I,\\
f^\phi &= 0.
\end{aligned}$$

Nonlinear damping terms can also be included, without substantially changing the results [19].

7.1 Unperturbed dynamics. The unperturbed system, obtained by putting $\epsilon = 0$ in (7.3), is an integrable Hamiltonian system, with two constants of motion: the unperturbed Hamiltonian H_0 and the action coordinate I. Recall that for the shaft, we have

$$\beta_1 > 0, \quad \beta_2 < 0, \quad \alpha_1 < 0, \quad \alpha_2 < 0, \quad \alpha_3 < 0,$$

which we assume to hold throughout the discussion below.

The first three equations of the unperturbed system,

$$\begin{aligned} \dot{x} &= \beta_1 \delta y + 32\alpha_1 y^3 + 8\alpha_2 I y, \\ \dot{y} &= x, \\ \dot{I} &= 0, \end{aligned} \tag{7.5}$$

are identical to the first three equations of the unforced undamped system (3.2). Thus, we have the phase portraits of Figure 2. In particular, when $\delta > 0$ for (7.5), we have equilibria

$$E_0: \qquad x = 0, \quad y = 0, \quad I = \text{constant}$$

on the I-axis, and additional pairs of equilibria

$$E_1^{\pm}: \qquad x = 0, \quad y = \pm\sqrt{\frac{\beta_1\delta + 8\alpha_2 I}{-32\alpha_1}}, \quad I = \text{constant}$$

that exist provided that

$$0 < I < \frac{\beta_1 \delta}{8|\alpha_2|}. \tag{7.6}$$

See Figure 2(b). When (7.6) holds, the equilibria E_0 are unstable (they are hyperbolic saddle points in the two-dimensional planes $\{(x, y, I) \mid I = \text{constant}\}$) while the equilibria $E_1^{\pm}$ are stable, and moreover there are pairs of orbits

$$x = x_{\pm}^h(t, I), \qquad y = y_{\pm}^h(t, I), \qquad I = \text{constant}$$

that are homoclinic to E_0, with

$$\lim_{t\to\pm\infty} x_{\pm}^h(t, I) = 0, \qquad \lim_{t\to\pm\infty} y_{\pm}^h(t, I) = 0,$$

in the planes $\{(x, y, I) \mid I = \text{constant}\}$. One can determine explicit expressions for these homoclinic orbits, see [19].

Now let us consider the last two equations of the unperturbed system. In particular, on the invariant plane

$$\Pi_0 = \{(x, y, I, \phi) \mid x = y = 0\}$$

we have for $\epsilon = 0$

$$\begin{aligned} \dot{I} &= 0, \\ \dot{\phi} &= \beta_2 \delta - \lambda \Lambda_1 + 4\alpha_3 I. \end{aligned} \tag{7.7}$$

Thus, all orbits on Π_0 are circles $\{I = \text{constant}\}$. In particular, for

$$I = I_0 = \frac{\lambda\Lambda_1 - \beta_2\delta}{4\alpha_3}$$

there is a circle of equilibria ($\dot{\phi} = 0$), or resonances, which we denote by $\mathcal{C}$. For a real physical system we must have $I_0 > 0$ in order for $\mathcal{C}$ to exist, and thus for the shaft, we have the condition

$$\delta < -\frac{\lambda\Lambda_1}{|\beta_2|} \tag{7.8}$$

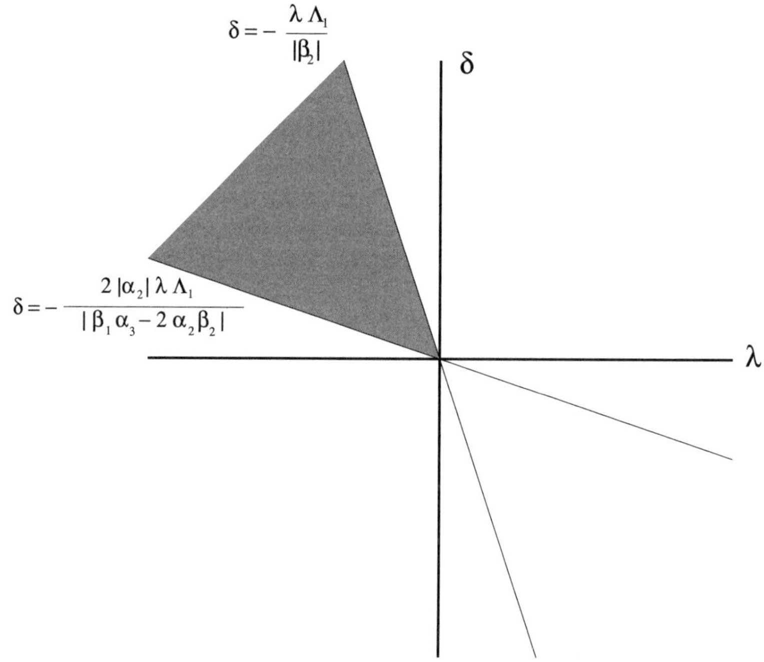

Figure 7 Region where resonant circle $\mathcal{C}$ and orbits homoclinic to $\mathcal{C}$ exist for the unperturbed system ($\epsilon = 0$).

for the existence of $\mathcal{C}$. In addition, the twist condition

$$D_I^2 H_0(\Pi_0)\big|_{I=I_0} = 4\alpha_3 \neq 0$$

is satisfied. The equilibria for (7.7) at $I = I_0$ correspond to periodic solutions with the same frequency as the forcing frequency.

If we substitute in the resonant value $I = I_0$ into the condition (7.6) for a homoclinic orbit to occur, we obtain (using the fact that $\alpha_3\beta_1 - 2\alpha_2\beta_2 < 0$ for the shaft)

$$\delta > -\frac{2\,|\alpha_2|\,\lambda\Lambda_1}{|\alpha_3\beta_1 - 2\alpha_2\beta_2|} \tag{7.9}$$

Combining this condition with the condition (7.8) for a resonant circle $\mathcal{C}$ to exist, we obtain a region in the $\delta\lambda$-plane where a resonant circle exists in the invariant manifold Π_0, and there is an orbit homoclinic to that resonant circle. The analysis in this section will apply to this region. A typical such region for the shaft is shown in in Figure 7.

Finally, for the unperturbed system we may calculate the phase shift of the angular coordinate ϕ. In the four-dimensional system (7.3) with $\epsilon = 0$, every equilibrium $(0, 0, I_0, \phi_0)$ on the resonant circle $\mathcal{C}$ is connected to another equilibrium $(0, 0, I_0, \phi_0 + \Delta\phi)$ on $\mathcal{C}$ by a pair of heteroclinic orbits

$$x_\pm^h(t, I_0), \quad y_\pm^h(t, I_0), \quad I(t, I_0) = I_0,$$

$$\phi(t, I_0, \phi_0) = \phi_0 + \int_{-\infty}^{t} \{\beta_2\delta - \lambda\Lambda_1 + 4\alpha_2[y_\pm^h(s; I_0)]^2 + 4\alpha_3 I_0\}\, ds.$$

Thus the global stable and unstable manifolds of Π_0 intersect, and the intersection contains these heteroclinic orbits. By using explicit expressions for $y_\pm^h(s;I_0)$, the phase shift in ϕ along either orbit can be calculated as

$$\begin{aligned}\Delta\phi &= \phi(+\infty, I_0, \phi_0) - \phi(-\infty, I_0, \phi_0)\\ &= \int_{-\infty}^{\infty} \{\beta_2\delta - \lambda\Lambda_1 + 4\alpha_2[y_\perp^h(s;I_0)]^2 + 4\alpha_3 I_0\}\,ds \\ &= \frac{\alpha_2}{2\alpha_1}\sqrt{\beta_1\delta + 8\alpha_2 I_0}.\end{aligned} \tag{7.10}$$

See [19] for details.

7.2 Perturbed dynamics. When $\epsilon > 0$ in (7.3), we have the perturbed system with small forcing and damping. We observe that Π_0 is still invariant when $\epsilon > 0$, and the theory of invariant manifolds tells us that the local stable and unstable manifolds of Π_0 persist. However, we need to study how the dynamics restricted to Π_0 may be changed by the perturbation, and also whether the global stable and unstable manifolds still intersect.

To study the dynamics on Π_0 near the resonant circle $\mathcal{C}$ when $\epsilon > 0$, we make the standard coordinate transformation

$$I = I_0 + \sqrt{\epsilon}\,\eta$$

and rescale time by $\tau = \sqrt{\epsilon}\,t$. Then (7.3) with $x = y = 0$ becomes

$$\begin{aligned}\eta' &= 2\sigma_4 I_0 \sin 2\phi + 2\hat{\zeta}\delta_2 I_0 + \sqrt{\epsilon}\,G(\eta,\phi) + \mathcal{O}(\epsilon),\\ \phi' &= 4\alpha_3\eta + \sqrt{\epsilon}\,F(\eta,\phi) + \mathcal{O}(\epsilon)\end{aligned} \tag{7.11}$$

where

$$G(\eta,\phi) = (2\sigma_4\sin 2\phi + 2\hat{\zeta}\delta_2)\eta \qquad F(\eta,\phi) = \sigma_4\cos 2\phi. \tag{7.12}$$

To leading order in the singular limit $\epsilon \to 0$, the system (7.11) is locally Hamiltonian:

$$\begin{aligned}\eta' &= 2\sigma_4 I_0\sin 2\phi + 2\hat{\zeta}\delta_2 I_0 = -D_\phi\mathcal{H},\\ \theta' &= 4\alpha_3\eta = D_\eta\mathcal{H},\end{aligned} \tag{7.13}$$

where

$$\mathcal{H} = 2\alpha_3\eta^2 + \sigma_4 I_0\cos 2\phi - 2\hat{\zeta}\delta_2 I_0\phi.$$

Because the leading-order system (7.13) is π-periodic in ϕ, we need only look at the dynamics for $\phi \in (-\pi/2, \pi/2]$. The system (7.13) has two equilibria: a centre

$$p_c^0 = (\eta_c, \phi_c) = \left(0, \frac{1}{2}\sin^{-1}\frac{\hat{\zeta}\delta_2}{\sigma_4}\right)$$

and a saddle

$$p_s^0 = (\eta_s, \phi_s) = \left(0, \frac{\pi}{2} - \frac{1}{2}\sin^{-1}\frac{\hat{\zeta}\delta_2}{\sigma_4}\right).$$

There is a homoclinic orbit connecting the saddle point p_s^0 to itself and the centre p_c^0 is the only equilibrium inside this homoclinic orbit. The value of ϕ at the "nose" of this homoclinic orbit, denoted by ϕ_n, can be determined by solving

$$\mathcal{H}(0,\phi_n) = \mathcal{H}(0,\phi_s)$$

for ϕ_n.

When $\epsilon > 0$ in (7.11), the hyperbolic saddle point of the leading-order system persists as a hyperbolic saddle point p_s^ϵ, while the centre of the leading-order system also persists as an equilibrium p_c^ϵ but becomes a hyperbolic sink when $\epsilon > 0$ (by Bendixson's criterion, see [19]). Although the homoclinic orbit of the leading-order system will not generally persist under perturbation, the homoclinic orbit itself is a good approximation to the basin of attraction of the sink p_c^ϵ when $\epsilon > 0$ is small.

To determine whether an orbit is homoclinic to the sink p_c^ϵ, we need to verify two conditions. First, we need to ensure that part of the unstable manifold of p_c^ϵ returns to the manifold Π_0. This requires the calculation of a Melnikov function. The second condition we need to check is the phase condition. Namely, we need to determine whether the orbit in the unstable manifold of p_c^ϵ, when it returns to the manifold Π_0, falls within the basin of attraction of the hyperbolic sink p_c^ϵ.

The Melnikov function is given in terms of the equations in standard form (7.4) by

$$M(I_0, \phi_c) = \int_{-\infty}^{\infty} \left\{ \langle D_{\mathbf{x}} H_0, J D_{\mathbf{x}} H_1 + f^{\mathbf{x}} \rangle + D_I H_0 \cdot \left(-D_\phi H_1 + f^I\right) \right\} dt,$$

where the angle brackets $\langle\, , \rangle$ denote the inner product in $\mathbb{R}^2$ and the terms in the integrand are evaluated along the orbit $\left(\mathbf{x}^h(t, I_0), I_0, \phi(t, I_0, \phi_c)\right)$. After some calculation we obtain [19]

$$M(I_0, \phi_c) = -\sigma_4 I_0 \left[\cos 2(\phi_c + \Delta\phi) - \cos 2\phi_c\right] + \hat{\zeta} K \Delta\phi$$

where

$$K = -\frac{2\delta_1(8\alpha_2 I_0 + \beta_1 \delta)}{12\alpha_2(1+\lambda)} + 2\delta_2 I_0$$

For the phase condition, it is sufficient to use the homoclinic orbit of the unperturbed leading-order system to approximate the basin. Thus the phase condition can be expressed as

$$\phi_s < \phi^\infty < \phi_n,$$

where $\phi^\infty = \phi_c + \Delta\phi$.

If there exist parameter values $(\delta, \lambda, \hat{\zeta})$ such that the Melnikov function has a simple zero with respect to the parameters and the phase condition is also satisfied, then for all sufficiently small $\epsilon > 0$, the forced damped system (7.3) in the subharmonic resonance case has a homoclinic orbit connecting the saddle-focus p_c^ϵ. Such orbits (sometimes called Šilnikov type homoclinic orbits) give rise to Smale horseshoes and chaotic dynamics in (7.3). However, (7.3) itself is an approximation of the forced damped system that neglects higher-order terms, some of which are nonautonomous and periodic. Thus (7.3) may be thought of as generating an *approximate* four-dimensional Poincaré map for the nonautonomous periodically forced system of equations that models the full parametrically excited damped system. If (7.3) has chaotic dynamics, it seems likely that the full nonautonomous system also has some complicated dynamics.

7.3 Physical interpretation of the global dynamics. For the rotating shaft, the dynamics of the system can be interpreted in terms of forward and backward whirling modes. As mentioned in the Appendix, a shaft may have two distinct motions–rotation and whirling. In a whirling motion, each section of the shaft undergoes an orbital motion about the centerline of the shaft. If the direction of the whirl is in the same direction as the rotation, relative to an inertial frame, then the whirl is called a *forward whirl,* while if the whirling motion is in the opposite direction as the rotation, relative to an inertial frame, then the whirl is called a *backward whirl.* We denote the whirling frequencies of these two motion as ω_f and ω_b respectively. Since we are considering the shaft in a rotating frame, the whirling frequencies will appear to be different, relative to the rotating frame. In particular, since the shaft is rotating with frequency Ω_0, the forward and backward whirling frequencies in the rotating frame are given by $\omega_f - \Omega_0$ and $\omega_b + \Omega_0$ respectively. Thus, if the rotation rate $\Omega_0 > \omega_f$, then in a rotating frame, the forward whirl will appear to be a backward whirl. The backward whirl will always appear to be backwards in the rotating frame, but with the increased frequency $\omega_b + \Omega_0$. The forward and backward whirling motions are shown in Figure 8. When the rotation rate exactly equals the forward whirl rate, $\Omega_0 = \omega_f$, then we have a *synchronous whirl.* A rotating observer will then see the forward whirl as a constant amplitude and constant phase displacement away from the trivial solution.

We are studying the rotating shaft near the case where a synchronous whirl occurs. Thus, in the rotating system, we have a frequency near zero. The first mode in our analysis, represented by (x, y) in (7.3), corresponds to the forward whirling mode of the shaft, while the second mode, represented by (I, ϕ) on the invariant manifold Π_0 corresponds to the backwards whirling mode of the shaft. The backwards whirling mode has frequency $\omega_f + \Omega_0 = \Lambda_1$, which is large compared to the near-zero frequency of the forward whirling mode. The trivial solution $(x, y, I) = (0, 0, 0)$ of (7.3) corresponds to no whirling in the shaft, i.e. a pure rotation. Motions on the invariant manifold Π_0 with $I > 0$ correspond to pure backward whirling motions. For $\delta > 0$ and I near $I_0 > 0$ we have the existence of a pair of spiral foci p_c^ϵ on Π_0 when $\epsilon > 0$ in (7.11),. These correspond to constant amplitude and phase backwards whirls. This motion, shown in Figure 9 in the original system coordinates q_1 and q_2, is a resonant backwards whirl that results from the parametric excitation, with frequency equal to the forcing frequency. For $\delta > 0$ and $0 < I < \beta_1\delta/(8\,|\alpha_2|)$, we also have homoclinic orbits in the xy-plane when $\epsilon = 0$. For the shaft, these orbits asymptotically approach the pure backward whirling motion in forwards and backwards time. In between these asymptotic limits, the motion of the shaft develops and then loses a component in the forward whirling mode, which can be seen in Figure 9. The pure backwards whirling motion is seen as a circular motion centered about $(q_1, q_2) = (0, 0)$. The motion of the shaft develops an offset and makes some oscillations about this offset before returning to backwards whirling motion. For $\epsilon > 0$ the Šilnikov type homoclinic orbit described above approaches the constant amplitude and phase backwards whirl p_c^ϵ in backwards time, develops a component in the forward whirling mode as time increases, and in forwards time loses the forwards whirling and approaches the constant amplitude and phase backwards whirl again, but approaches slowly near the manifold Π_0 of backwards whirling motions. Any associated chaotic motion spends most of the time as nonconstant backwards whirling motion, with relatively brief and seemingly random (in time) excursions developing and then losing components in the forward whirling mode.

8 Conclusions

We can now summarize the results from this paper. When we add forcing at the subharmonic resonance frequency, we add nonautonomous terms to the equations

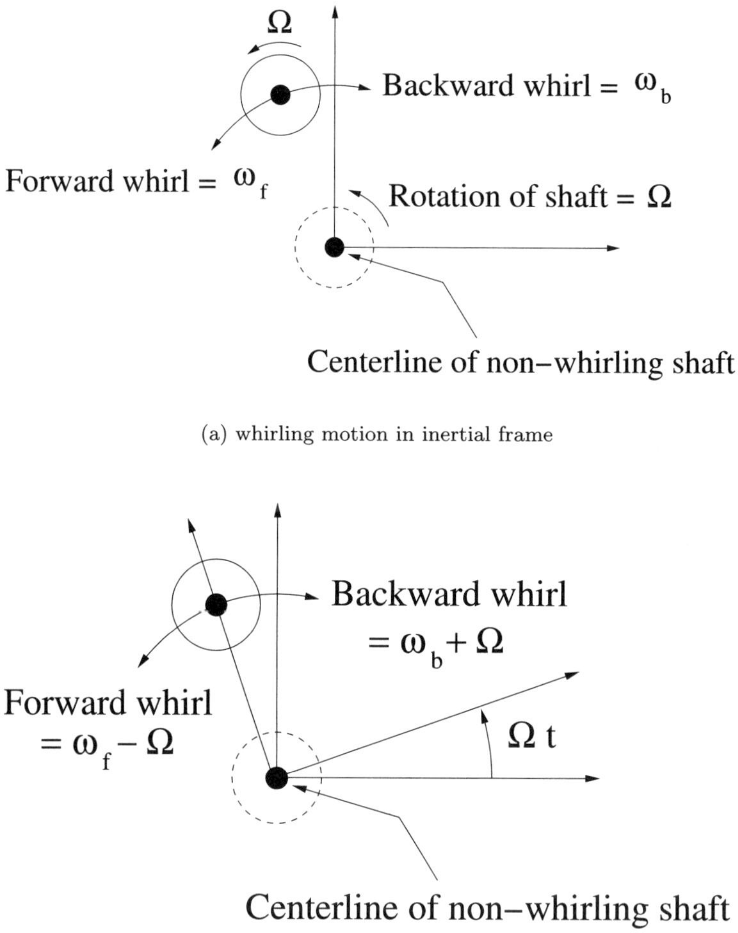

(a) whirling motion in inertial frame

(b) Whirling motion in rotating frame

Figure 8 Whirling motion of shaft in inertial and rotating frames.

of motion. To study this system, we make a time-dependent transformation that eliminates the explicit time dependence from the low-order terms in the equations of motion. This transformation also moves the eigenvalue pair that was far from zero on the imaginary axis to the neighborhood of zero. If we had made this transformation for the unforced system, these eigenvalues would simply pass at zero as δ varied, due to the phase-shift S^1-symmetry of the second mode. However, the addition of forcing breaks this symmetry, and hence the second mode can become unstable here. In fact, this behavior of the second mode is similar to the behavior of our gyroscopic system as a whole: symmetry-breaking causes an instability in the system, followed by a re-stabilization. Of course, the system cannot be re-stabilized if the first mode has already become unstable.

For the shaft, each of the primary bifurcations is subcritical. The behavior of the non-trivial solutions for the shaft is such that, for the forced system there may

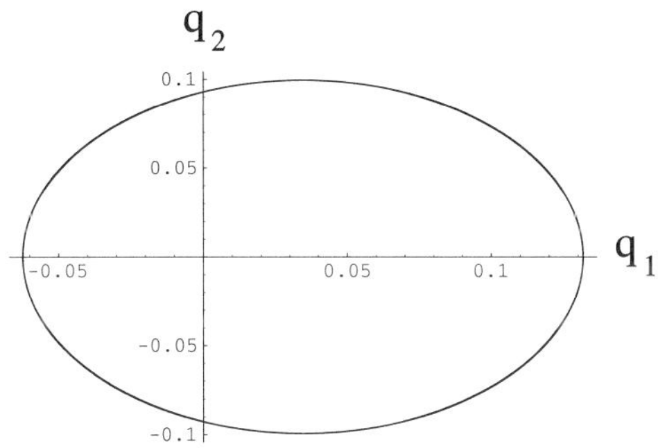

(a) Center fixed point

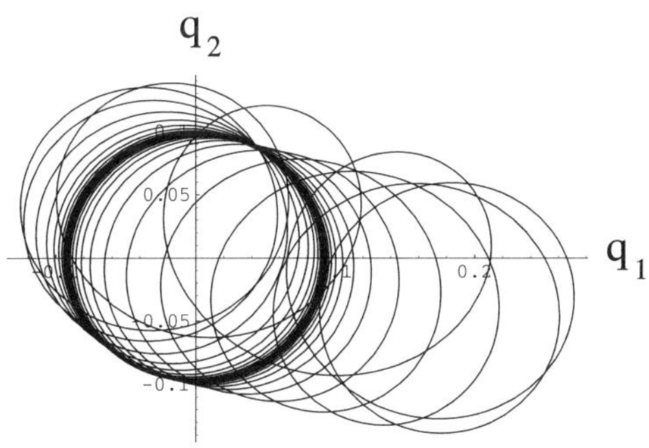

(b) Homoclinic orbit

Figure 9 Global motions of the rotating shaft.

be 0, 1, or 2 multi-mode solutions connecting the primary bifurcation branches, depending on the direction and size of the detuning.

Finally in this paper, we were able to identify conditions under which a Šilnikov type homoclinic orbit exists for the truncated, autonomous normal form of the forced damped system. Under such conditions, the full, nonautonomous system likely undergoes complicated dynamics.

Acknowledgments. The second author would like to acknowledge the support of National Science Foundation under grant number CMS 03-01412 and DMS 05-04581. Any opinions, findings, and conclusions or recommendations expressed in this paper are those of the authors and do not necessarily reflect the views of the National Science Foundation. The third author acknowledges the support of the Natural Sciences and Engineering Research Council of Canada.

Appendix: Equations of motion for rotating shafts

The dynamics of rotating systems such as shafts have been studied extensively in the literature (Dimentberg [4], Tondl [27], Bolotin [3], Mettler [20], Shaw and Shaw [25]). More recently, Shaw and Shaw [26] have studied the effects of

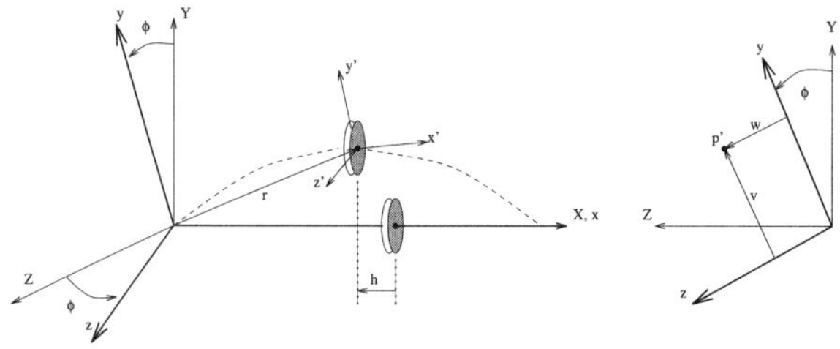

Figure 10 Stationary, rotating and body-fixed reference frames.

damping and symmetry-breaking on perfectly symmetric shafts. Nagata and Namachchivaya [22] studied the local and global bifurcations of damped gyroscopic systems, using a rotating shaft as their main example. We extend the derivation of Nagata and Sri Namachchivaya [22] by including time-varying rotation rate as in McDonald [16].

Consider a shaft of length L, with mass per unit length m, and flexural rigidities EI_{yy} and EI_{zz} parallel to the y- and z-axes, respectively (E is Young's modulus, while I_{yy} and I_{zz} are the lateral area moments of inertia of the shaft cross section with respect to the y-axis and the z-axis, respectively). We assume $I_{yy} < I_{zz}$. The equations governing the transverse vibrations of a rotating shaft subjected to an axial load are derived for a shaft of length L and constant cross-sectional area A. Consider Figure 10 depicting the stationary (X, Y, Z), rotating (x, y, z) and body-fixed (x', y', z') coordinate systems. The vector $\vec{r}$ locates the center-line p' at a particular cross-section. This model also assumes that the shaft is rotating in torque-free bearings (e.g. ideal superconducting or magnetic bearings). The latter restriction may be relaxed by considering additional terms describing the torque and resulting twist in the shaft. For clarity in presenting the method of analysis, the effect of torque has been omitted.

The nonlinearities in the final equations of motion stem from the geometry of the mechanical problem: mid-plane stretching and large curvature. The equations of motion are derived here using the energy method. The potential energy, including the combined effects of mid-plane stretching and bending, and the kinetic energy are calculated and combined to form the Lagrangian. Hamilton's Principle of Least Action yields the partial differential equations of motion. Adding the dissipation effects we obtain the equation of motion for the transverse motion as

$$
\begin{aligned}
0 \;=\;& m\left[\ddot{v} - \dot{\Omega} w - 2\Omega\dot{w} - \Omega^2 v\right] + \mu_e \dot{v} - \mu_e \Omega w + EI_{yy} v^{iv} + \mu_i EI_{yy} \dot{v}^{iv} \\
& -\frac{EA}{2L}\left\{\left(1 + \mu_i \frac{d}{dt}\right)\int_0^L \left((v')^2 + (w')^2\right)\,\mathrm{ds}\right\} v'' \\
& +EI_{yy}\left[v^{iv}\left((v')^2 + (w')^2\right) + v''\left((v'')^2 + 2(w'')^2\right)\right. \\
& \left. +4v'''\left(v'v'' + w'w''\right) + 2w'v''w'''\right] \\
& -EI_{zz}\left[v''(w'')^2 + 2v'w''w'''\right]
\end{aligned} \tag{8.1}
$$

$$
\begin{aligned}
0 \;=\;& m\left[\ddot{w}+\dot{\Omega}v+2\Omega\dot{v}-\Omega^2 w\right]+\mu_e\dot{w}+\mu_e\Omega v+EI_{zz}w^{iv}+\mu_i EI_{zz}\dot{w}^{iv} \\
&-\frac{EA}{2L}\left\{\left(1+\mu_i\frac{d}{dt}\right)\int_0^L\left((v')^2+(w')^2\right)\,\mathrm{ds}\right\}w'' \\
&+EI_{zz}\left[w^{iv}\left((v')^2+(w')^2\right)+w''\left(2(v'')^2+(w'')^2\right)\right. \\
&\left.+4w'''\left(v'v''+w'w''\right)+2v'w''v'''\right] \\
&-EI_{yy}\left[(v'')^2w''+2v'''v''w'\right]
\end{aligned}
\tag{8.2}
$$

In our equations for transverse displacements v, w, we added internal damping terms μ_i into the axial stress terms and the linear bending terms. We also add in external damping terms μ_e, due to friction between the shaft and its supports. We non-dimensionalize the equations to obtain a simpler form, using the substitutions

$$
\tau=\sqrt{\frac{EI_{yy}}{mL^4}}\,t=\gamma t, \qquad \overline{v}=\frac{v}{L}, \qquad \overline{w}=\frac{w}{L}, \qquad \overline{s}=\frac{s}{L}
$$

$$
\overline{\mu}_e=\frac{\mu_e}{m\gamma}, \qquad \overline{\mu}_i=\mu_i\gamma, \qquad \alpha=\frac{AL^2}{2I_{yy}}, \qquad \overline{\Omega}=\frac{\Omega}{\gamma}
$$

and divide by the common factor $m\gamma^2 L=\frac{EI_{yy}}{L^3}$ to obtain the non-dimensionalized PDEs (after eliminating the overbars for notational simplicity)

$$
\begin{aligned}
0 \;=\;& \ddot{v}-\dot{\Omega}w-2\Omega\dot{w}-\Omega^2 v+\mu_e\dot{v}-\mu_e\Omega w+v^{iv}+\mu_i\dot{v}^{iv} \\
&-\alpha\left\{\left(1+\mu_i\frac{d}{dt}\right)\int_0^1\left((v')^2+(w')^2\right)\,\mathrm{ds}\right\}v'' \\
&+v^{iv}\left((v')^2+(w')^2\right)+v''\left((v'')^2+2(w'')^2\right) \\
&+4v'''\left(v''\,v'+w''\,w'\right)+2w'''\,v''\,w' \\
&-\frac{I_{zz}}{I_{yy}}\left\{v''(w'')^2+2w'''\,w''\,v'\right\}
\end{aligned}
\tag{8.3}
$$

$$
\begin{aligned}
0 \;=\;& \ddot{w}+\dot{\Omega}v+2\Omega\dot{v}-\Omega^2 w+\mu_e\dot{w}+\mu_e\Omega v+\frac{I_{zz}}{I_{yy}}w^{iv}+\frac{I_{zz}}{I_{yy}}\mu_i\dot{w}^{iv} \\
&-\alpha\left\{\left(1+\mu_i\frac{d}{dt}\right)\int_0^1\left((v')^2+(w')^2\right)\,\mathrm{ds}\right\}w'' \\
&+\frac{I_{zz}}{I_{yy}}\left\{w^{iv}\left((v')^2+(w')^2\right)+w''\left(2(v'')^2+(w'')^2\right)\right. \\
&\left.+4w'''\left(v''v'+w''w'\right)+2v'''w''v'\right\} \\
&-\left\{(v'')^2w''+2v'''v''w'\right\}
\end{aligned}
\tag{8.4}
$$

Applying a modal approximation reduces the system of partial differential equations to ordinary differential equations governing the amplitude of vibration in each of the two principal directions. Thus substituting for $v(s,t)=q_1(t)\phi(s)$ and $w(s,t)=q_2(t)\phi(s)$ respectively in the (y', z') directions, where $\phi(s)=\sin\pi s$ describes the mode shape of the first mode, the equation for the amplitudes $q_1(t)$ and

$q_2(t)$ are given by: (see [16] for more details):

$$\begin{aligned} 0 &= \ddot{q}_1 - \dot{\Omega} q_2 - 2\Omega \dot{q}_2 - \Omega^2 q_1 + \mu_e \dot{q}_1 - \mu_e \Omega q_2 + (q_1 + \mu_i \dot{q}_1)\,\pi^4 \\ &\quad + \frac{\partial U}{\partial q_1} + \mu_i \alpha \pi^4 q_1 (2q_1\dot{q}_1 + 2q_2\dot{q}_2) \\ 0 &= \ddot{q}_2 + \dot{\Omega} q_1 + 2\Omega \dot{q}_1 - \Omega^2 q_2 + \mu_e \dot{q}_2 + \mu_e \Omega q_1 + \frac{I_{zz}}{I_{yy}} (q_2 + \mu_i \dot{q}_2)\,\pi^4 \\ &\quad + \frac{\partial U}{\partial q_2} + \mu_i \alpha \pi^4 q_2 (2q_1\dot{q}_1 + 2q_2\dot{q}_2) \end{aligned}$$

where the nonlinear conservative terms have been written in terms of the potential

$$U = \frac{\alpha\pi^4}{4}\left(q_1^2 + q_2^2\right)^2 + \frac{\pi^6}{4}\left(q_1^4 + \left(1 + \frac{I_{zz}}{I_{yy}}\right) q_1^2 q_2^2 + \frac{I_{zz}}{I_{yy}} q_2^4\right) \tag{8.5}$$

In vector form, these equations can be written

$$\begin{aligned} 0 =& \ddot{\mathbf{q}} + 2G(t)\dot{\mathbf{q}} + \dot{G}(t)\mathbf{q} + K(t)\mathbf{q} + \mu_e \dot{\mathbf{q}} + \mu_e G(t)\mathbf{q} + \mu_i \Lambda \dot{\mathbf{q}} \\ &+ \nabla U(\mathbf{q}) + F(\mathbf{q}, \dot{\mathbf{q}}) \end{aligned} \tag{8.6}$$

where

$$\mathbf{q} = \begin{bmatrix} q_1 \\ q_2 \end{bmatrix}, \qquad G(t) = \begin{bmatrix} 0 & -\Omega(t) \\ \Omega(t) & 0 \end{bmatrix},$$

$$K(t) = \begin{bmatrix} \pi^4 - \Omega^2 & 0 \\ 0 & \frac{I_{zz}}{I_{yy}}\pi^4 - \Omega^2 \end{bmatrix} \qquad \Lambda = \begin{bmatrix} \mu_i \pi^4 & 0 \\ 0 & \frac{I_{zz}}{I_{yy}}\mu_i \pi^4 \end{bmatrix}$$

G represents the gyroscopic terms, K represents stiffness terms, Λ represents linear internal damping terms, $\mu_e \dot{\mathbf{q}} + \mu_e G(t)\mathbf{q}$ are the external linear damping terms, $\nabla U(\mathbf{q})$ represents nonlinear conservative terms, and

$$F(\mathbf{q}, \dot{\mathbf{q}}) = 2\mu_i \alpha \pi^4 (q_1\dot{q}_1 + q_2\dot{q}_2) \begin{bmatrix} q_1 \\ q_2 \end{bmatrix}$$

represents nonlinear internal damping terms. Note that for the non-rotating system, the natural frequencies of the shaft are given by $\bar{\omega}_1^2 = \pi^4$ and $\bar{\omega}_2^2 = \frac{I_{zz}}{I_{yy}}\pi^4$, while for the rotating system, the natural frequencies are given by $\omega_1^2 = \pi^4 - \Omega^2$ and $\omega_2^2 = \frac{I_{zz}}{I_{yy}}\pi^4 - \Omega^2$. Thus, the rotation of the shaft has the effect of decreasing its effective stiffness. If the rotation rate becomes large enough, this stiffness will go to zero, and the rotating shaft will exhibit whirling motion.

Normal form coefficients. We shall be very brief here, concentrating on the form of the coefficients in the Hamiltonian normal forms (2.8) and (2.15). Details of the calculations, which are straightforward but lengthy, can be found in [16]. We can determine the coefficients of the nonlinear terms as

$$\alpha_1 = -\frac{9\alpha\pi^{12}}{2\Lambda_1^4}, \quad \alpha_2 = -\frac{6\alpha\pi^8\left(\Lambda_1^2 + 3\bar{\omega}_1^2\right)}{\Lambda_1^5}, \quad \alpha_3 = -\frac{\alpha\pi^4\left(3\Lambda_1^4 + 2\Lambda_1^2\bar{\omega}_1^2 + 3\bar{\omega}_1^4\right)}{\Lambda_1^6},$$

where $\Lambda_1^2 = 3\bar{\omega}_1^2 + \bar{\omega}_2^2$. Since the the physical parameter $\alpha = AL^2/(2I_{yy})$ is positive, we can see that each of the coefficients α_1, α_2, α_3 is negative. The coefficients of the autonomous unfolding terms can be determined as

$$\beta_1 = \frac{4\pi^2(\beta - 1)}{(3 + \beta)}, \qquad \beta_2 = -\frac{2(5 + 3\beta)}{(3 + \beta)^{\frac{3}{2}}},$$

where

$$\beta = \frac{\bar{\omega}_2^2}{\bar{\omega}_1^2} > 1.$$

It is clear that β_1 is positive and β_2 is negative. The coefficients of the forcing terms can be determined as

$$\sigma_1 = -\frac{\pi\,(1+i)\,(\beta-1)^{\frac{3}{2}}}{2\sqrt{2}\,(3+\beta)^{\frac{5}{4}}}, \qquad \sigma_4 = \frac{(\beta-1)}{(3+\beta)^{\frac{3}{2}}}.$$

We can see that both the real and imaginary parts of σ_1 are negative, and σ_4 is real and positive. Finally, we can find the following damping normal form coefficients for external and internal damping respectively,

$$\delta_1^e = -\frac{1}{2}, \qquad \delta_2^e = -\frac{1}{2}, \qquad \delta_3^e = 0,$$

$$\delta_1^i = -\frac{\pi^4\left(\bar{\omega}_2^2 - \bar{\omega}_1^2\right)}{2\Lambda_1^2} \qquad \delta_2^i = -\frac{\pi^4}{2}\left(\beta + \frac{4\bar{\omega}_1^2}{\Lambda_1^2}\right), \qquad \delta_3^i = 0,$$

and we write $\delta_1 = \delta_1^i + \delta_1^e$, $\delta_2 = \delta_2^i + \delta_2^e$, and $\delta_3 = \delta_3^i + \delta_3^e$.

References

[1] S. T. Ariaratnam and N. Sri Namachchivaya. Periodically perturbed linear gyroscopic systems. *Journal of Structural Mechanics*, 14(2):127–151, 1986.

[2] S. T. Ariaratnam and N. Sri Namachchivaya. Periodically perturbed nonlinear gyroscopic systems. *Journal of Structural Mechanics*, 14(2):153–175, 1986.

[3] V. V. Bolotin. *The Dynamic Stability of Elastic Systems*. Holden-Day, San Francisco, 1964.

[4] F. M. Dimentberg. *Flexural Vibrations of Rotating Shafts*. Butterworth, London, 1961.

[5] Z. Feng and P. R. Sethna. Global bifurcations in the motion of parametrically excited thin plates. *Nonlinear Dynamics*, 4:398–408, 1993.

[6] J. Guckenheimer and P. Holmes. *Nonlinear Oscillations, Dynamical Systems, and Bifurcations of Vector Fields*. Springer-Verlag, New York, 1983.

[7] J. Guckenheimer and A Mahalov. Instability induced by symmetry reduction. *Physical Review Letters*, 68(15):2257–2260, 1992.

[8] G. Haller. Multi-dimensional homoclinic jumping and the discretized NLS equation. *Comm. Math. Phys.*, 193:1–46, 1998.

[9] G. Haller and S. Wiggins. Multi-pulse jumping orbits and homoclinic trees in a modal truncation of the damped-forced nonlinear Schrödinger equation. *Physica D*, 85:311–347, 1995.

[10] G. Haller. *Chaos near resonance*. Springer-Verlag, New York, 1999.

[11] G. Kovačič and S. Wiggins. Orbits homoclinic to resonances, with an application to chaos in a model of the forced and damped sine-gordon equation. *Physica D*, 57:185–225, 1992.

[12] G. Kovačič and T. A. Wettergren. Homoclinic orbits in the dynamics of resonantly driven coupled pendula. *ZAMP*, 47:221–264, 1996.

[13] M. G. Krein. A generalization of several investigations of A. M. Lyapunov on linear differential equations with periodic coefficients. *Dokl. Akad. Nauk. SSSR*, 73:445–448, 1950.

[14] N. Malhotra and N. Sri Namachchivaya. Global bifurcations in externally excited two-degrees-of-freedom nonlinear systems. *Nonlinear Dynamics*, 8:85–109, 1995.

[15] N. Malhotra, N. Sri Namachchivaya, and R. J. McDonald. Single and multi-pulse orbits in the motion of flexible spinning discs. *Journal of Nonlinear Science*, 12(1):1-26 2002.

[16] R. McDonald. *Bifurcations of Parametrically Excited Gyroscopic Systems Near a* 0 : 1 *Resonance Ph.D. Thesis*, University of Illinois at Urbana–Champaign, 2000.

[17] R. McDonald, J.A. Murdoch and N. Sri Namachchivaya. Normal forms for nonlinear Hamiltonian systems with weak periodic perturbations. *Journal of Dynamics and Stability of Systems*, 14:187–211, 1999.

[18] R. J. McDonald and N. Sri Namachchivaya. Pipes conveying pulsating fluid near a 0 : 1 resonance: local bifurcations. *Journal of Fluids and Structures*, 21:629–664, 2005.

[19] R. J. McDonald and N. Sri Namachchivaya. Pipes conveying pulsating fluid near a 0 : 1 resonance: global bifurcations. *Journal of Fluids and Structures*, 21:665–687, 2005.

[20] E. Mettler. Combination resonances in mechanical systems under harmonic excitation. In *Proceedings of the Fourth Conference on Nonlinear Oscillations*, pages 51–70, Prague, 1968. Academia.

[21] K. R. Meyer and G. R. Hall. *Introduction to Hamiltonian Dynamical Systems and the N-Body Problem.* Springer-Verlag, New York, 1992.

[22] W. Nagata and N. Sri Namachchivaya. Bifurcations in gyroscopic systems with an application to rotating shafts. *Proceedings of the Royal Society: Series A*, 454:543–585, 1998.

[23] A. H. Nayfeh and D. T. Mook. Energy transfer from high-frequency to low-frequency modes in structures. *Trans. ASME*, 186:186–195, 1995.

[24] S. A. Nayfeh and A. H. Nayfeh. Energy transfer from high- to low-frequency modes in a flexible structure via modulation. *J. Vibr. Acoust.*, 116:203–207, 1994.

[25] J. Shaw and S. W. Shaw. Instabilities and bifurcations in a rotating shaft. *Journal of Sound and Vibration*, 132(2):227–244, 1989.

[26] J. Shaw and S. W. Shaw. Non-linear resonance of an unbalanced rotating shaft with internal damping. *Journal of Sound and Vibration*, 147(3):435–451, 1991.

[27] A. Tondl. *Some Problems of Rotor Dynamics.* Publishing House of the Czechoslavak Academy of Science, Prague, 1965.

Fields Institute Communications
Volume **49**, 2006

High Dimensional Data Clustering from a Dynamical Systems Point of View

Jianhong Wu
Laboratory for Industrial and Applied Mathematics
Department of Mathematics and Statistics
York University
Toronto, Ontario, Canada, M3J 1P3
wujh@mathstat.yorku.ca

This paper is dedicated to the 60th birthday of Professor Bill Langford.

Abstract. We provide a short survey of some recent progress towards subspace clustering of high dimensional data (both numerical and categorical) from the viewpoint of dynamical systems. The novel idea behind the dynamical system based approach is the possible connection between the centers of different clusters in a given data set and the local attractors of a dynamical system to be constructed, and the connection between the clustering criteria not given *a priori* and the boundaries of domains of attraction of the aforementioned local attractors. The essential difficulty is, for a given data set, that one needs to construct the dynamical system and to discover the structures of the clusters (including the subspaces where these clusters are formed) simultaneously. We will present the related neural network architectures, the modeling dynamical systems, the qualitative analysis of the dynamical systems, the resulting clustering algorithms and their performance analysis and applications (to gene expression data analysis and to neural signal spike trains clustering), the connection to the bifurcation and pattern formation theory, and the mathematical and neurophysiological foundation in terms of adaptive delay in signal transmissions.

2000 *Mathematics Subject Classification.* Primary 34K10, 91C20, 65K10; Secondary 37N99, 37N40.

Research was partially supported by Natural Sciences and Engineering Research Council of Canada, by Canada Research Chairs Program, and by Network of Centers of Excellence Program: Mathematics for Information Technology and Complex Systems.

1 Introduction

This paper provides a short survey of the theory and applications of a recently developed Projective Adaptive Resonance Theory (PART) for pattern recognition and clustering in high dimensional data (both numerical and categorical) sets.

The focus is on the philosophical (and intuitive) aspect of the construction of the PART neural network architecture and the emphasis is on the connection between clustering as a cognitive and adaptive process and nonlinear dynamics as an evolutional and feedback process: the intuitive and intimate connections between the clusters of the data set and stable patterns, and between clustering criteria and the boundaries of domains of attraction of stable patterns of the constructed neural network dynamical system. The challenge lies in the nature of the task of unsupervised clustering: the number of clusters, the clustering criteria and the underlying dynamical systems all have to be discovered simultaneously. Adding to the difficulty is the high-dimensionality of the data set that requires to locate a cluster in a certain (unknown *a-priori*) subspace and hence a mechanism (selective output signal), that should be regarded as an inverse process of reduction of dynamics on invariant subspaces, becomes necessary.

A mathematical formulation of the aforementioned philosophical consideration takes the form of a large scale singularly perturbed nonlinear dynamical system with multi-time scales and adaptive time delays, and the construction of such a system is also guided by some recent biological and neurophysiological findings as well as some established learning laws. The full description of the transient and long time behaviors of the nonlinear dynamical system can be obtained for some commonly used nonlinear input-output functions, and the algorithm based on the nonlinear dynamical behaviors of PART seems to work quite well for synthetic data and for data arising from application problems.

Connection between the clustering process of a modified PART (IMC-PART) and the dynamics of a discrete map (iterations of a mapping in the data space) is also discussed, where the distance parameter becomes a natural bifurcation parameter: splitting a cluster is then naturally related to bifurcation theory in dynamical systems.

We also provide a short description of some recent applications of PART algorithm to clustering neural spiking trains via transient behaviors, to gene expression profiling, and to a Canadian Bank branch data for performance analysis, in order to show the usefulness of the algorithm. The paper also provides some comparisons with other existing projective clustering algorithms, and concludes with some discussions on remaining challenges.

2 Data clustering viewed as an inverse process of pattern formation

2.1 Data clustering and pattern recognition. Data clustering is one of the most common cognitive tasks performed by our central nervous system regularly, and is also an essential step in most data mining processes.

The main goal of clustering is to reorganize subsets of data points into groups (clusters), aiming at discovering structure which is hidden in a data set.

Vaguely speaking, for a given data set in a multidimensional space, the purpose of data clustering is to find a partition of the data set into groups (*clusters*) such that points within the same cluster are similar but points from different clusters are dissimilar according to certain *similarity criteria*. Unfortunately, most data clustering processes are unsupervised: the number of clusters and the similarity criteria used for the partition are unknown in advance.

Data clustering arises from every aspect of our daily life and from many important applications such as information retrieval, micro-array data analysis, online commerce, pattern recognition and image segmentation. Data clustering has also been intensively investigated for a long time, the readers are refereed to ([26], [27], [32], [37], [38], [48]) for some discussions and related references.

2.2 Pattern formation vs pattern recognition. We now describe our approach based on the idea that treats the data clustering problem from the view point of nonlinear dynamical systems, in the hope to construct a dynamical system to automatically and adaptively identify patterns hidden in a given data set. More precisely, we reformulate the data clustering problem as follows:

Data Clustering as an Inverse Problem of Pattern Formation: Given a data set D in R^m that can be decomposed as the union of disjoint subsets $A_1, \cdots, A_n$ such that points in each A_i are similar and points in different subsets are dissimilar according to some criteria expressed in terms of several surfaces $F(x_1, \cdots, x_m) = 0$ with $F : R^m \to R^l$. Construct a dynamical system so that

(G1) each subset A_i corresponds to a local attractor; and
(G2) each surface corresponds to a part of the boundaries of the domains of attractions of local attractors.

Recall that for a given dynamical system, identifying all possible stable patterns (local attractors) and determining their domains of attraction to see how these patterns are connected together to form a global attractor is the task of pattern formation. It is exactly in this sense that we view data clustering as an inverse process of pattern formation.

Figure 1 shows the essential idea in this viewpoint and philosophy. Here the data set consists of two groups (clusters), separated by a curve that gives the clustering criterion qualitatively. The red dots denote the cluster centers. Our goal is to construct a dynamical system so that the two cluster centers coincide with two stable equilibria of the dynamical system and the curve coincides with the boundary of the domains of attraction of the two stable equilibria. Note that there must be a unstable equilibrium (saddle point) and the curve is also the stable manifold of this saddle point.

There are several advantages of this approach, if successful. First of all, this method generates the centers of clusters (and numbers of clusters) automatically once the dynamical system is constructed from a given data set by simply calculating the stable equilibria (and this also gives the statistical mean/feature of the clusters). Secondly, we can qualitatively describe the clustering criteria as these coincide with the boundaries of domains of attraction of stable equilibria/attractors and these are usually some parts of the invariant stable manifolds of some unstable equilibria/solutions of the dynamical system, though in practice, calculating these invariant manifolds is by no means trivial. Thirdly, once the learning or the unsupervised clustering of a given data/sample set is complete, assigning a new data point to a cluster is straightforward from a dynamical system point of view: all we need to do is to determine whether the new point belongs to the domain of attraction of a particular stable equilibrium (the center of a cluster), and this can be achieved by simply following the trajectory starting from this data point and to see where it converges.

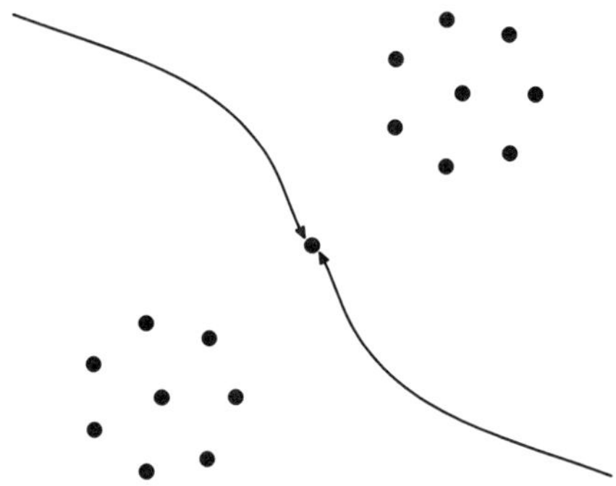

Figure 1 A data set consisting of two clusters, separated by a curve that gives the clustering criterion qualitatively. The red dots denote the cluster centers, and they coincide with two stable equilibria of a dynamical system (to be constructed) and the curve coincides with the boundary of the domains of attraction of two stable equilibria, this curve is also the stable manifold of a saddle point.

A question arises very naturally. Since many different types of nonlinear dynamical systems (including infinite dimensional dynamical systems such as partial differential equations and delay differential equations) share the same dynamical behaviors exhibited in Figure 1, that can be exhibited by a simple 2-dimensional system, what is the minimal size and structure of a nonlinear dynamical system required to identify the two clusters hidden in the given data set? This question seems to be well addressed in the celebrated Adaptive Resonance Theory developed by Carpenter and Grossberg, as will be discussed in next section.

There is another critical issue due to the high dimensionality of data sets generated by the modern technologies in data collection and processing. Coincidentally, this is related to another fundamental aspect of nonlinear dynamics: attractors imbedded in low dimensional manifolds of a high or even infinite dimensional dynamical system and restriction of dynamics on such low dimensional manifolds (center manifold reduction in a local sense, inertial manifold reduction globally, for example) captures the overall dynamical behaviors. Unfortunately, as will be discussed in next subsection, the stable patterns, their domains of attractions and these low dimensional manifolds have to be discovered simultaneously.

2.3 Feasibility-reliability dilemma. Data clustering has been a challenging task for scientific community for a long time, and this problem becomes increasingly important and difficult in today's information age due to the large volume and high dimensionality of data sets. High dimensional data sets arise often in a variety of contexts such as text/web mining, bioinformatics, imaging for diagnostics and surveillance, astronomy and remote sensing ([32]). Many technical challenges associated with the high dimensionality render traditional clustering methods ineffective because of the inherent sparsity of data. In such data sets, for any given pair of points there exist at least a few dimensions on which the points are far apart from one another. For example, for a data set that obeys the Gaussian distribution $N(\mu; \sigma^2 I_n)$, we must pick up at least 2^n random points from the data set in order to obtain just a few points which are at distance less than $\frac{1}{2}\sigma\sqrt{n}$ from the center. In addition, recent theoretical results ([8]) have shown that in a high dimensional space, the distance between every pair of points is almost the same for a wide variety of data distributions and distance functions. Therefore, it is not feasible to talk about proximity or clustering in the original full space of all dimensions (This is well known as the *curse of dimensionality*). Consequently, a clustering algorithm is often preceded by feature selection whose goal is to find the particular dimensions on which points in the data set are correlated. Unfortunately, a feature selection procedure requires picking up certain dimensions in advance without precise knowledge which dimensions are important, and this can lead to a significant loss of information. There are also many dimension reduction techniques such as PCA(Principal Component Analysis) [52] and Karhunen-Loève Transformation [1] developed in order to reduce the dimensionality before clustering. Unfortunately, such dimension reduction techniques share the same information loss problem as feature selection techniques. As a result, the clustering results preceded by a feature selection or dimension reduction procedure may not be reliable. This reliability problem becomes even worse in some data mining applications where some points are correlated with respect to a given set of dimensions and others are correlated with respect to different sets of dimensions, and therefore it may not always be possible to prune off dimensions without, at the same time, failing to find all interesting features and clusters.

In summary, we are facing a feasibility-reliability dilemma in clustering high dimensional data sets: on the one hand it is feasible to find clusters only in lower dimensional subspaces due to the sparsity of the data in full space, on the other hand the clustering results by pruning off dimensions in advance are not reliable due to the loss of information. This dilemma is illustrated in Figure 2, where a data set consists of three groups. Within a group, all data points are very close to each other with respect to two variables but these points also spread out along the third direction. Therefore, in searching for clusters in the full space, one concludes that there is either no cluster, or there are 4 clusters (those points close to the origin form one cluster, and the remaining points form three different clusters).

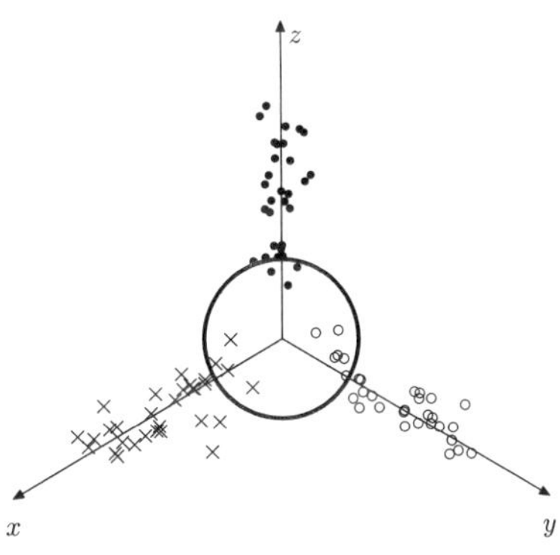

Figure 2 A data set with three groups, and searching for clusters in full 3-dimensional space will result in a false conclusion of no cluster or 4 clusters.

2.4 Projected/subspace clustering. This feasibility-reliability dilemma motivated the concept of *projected (subspace) clustering* ([2]) whose central goal is to find *projected (subspace) clusters*, each of which consists of a subset C_i of data points together with a subset D_i of dimensions such that the points in C_i are closely correlated in the subspace of dimensions D_i. The idea here is to find desired clusters based not only on points but also on dimensions. For example, in Figure 2 we have three projected clusters: the red dots together with the (x, y)-plane form one projected cluster, so do the green dots with the (z, y)-plane and the black dots with the (x, z)-plane.

CLIQUE [1] seems to be the first algorithm designed for automatically discovering clusters in different subspaces, using a density-based approach [37], [22], [44]. The idea is to partition every dimension into several intervals of equal length so that for a given subspace, a cross product of such intervals (one from each of dimensions of this subspace) gives a unit in the subspace. A unit is said to be dense if the fraction of total data points contained in the unit is above a density threshold. Then a maximal set of connected dense units in a given subspace forms a cluster in the subspace. Since a dense unit of k-dimensional subspace is also dense when projected into a $(k-1)$-dimensional subspace, CLIQUE first finds all dense units in 1-dimensional subspace, then works from lower dimensional subspaces to higher ones to discover dense clusters in all subspaces. The time complexity of CLIQUE is therefore exponential in the highest dimensionality of any dense unit, and this limits its applications.

A fast subspace clustering algorithm PROCLUS was proposed by Aggarwal et al in 1999 [2]. PROCLUS uses a medoids-based optimization approach [39], [43] and combines the greedy technique and a locality analysis to find the set of dimensions associated with each medoid (a medoid is a representative object in a cluster to serve as the surrogate center for the cluster). The clustering quality (the criterion function to be optimized) is evaluated as the average Manhattan segmental distance from all points to their corresponding cluster centers. Simulations on synthetic data sets show that PROCLUS performed better than CLIQUE in terms of quality and running time. However, a problem with this algorithm is how to determine its two input parameters: the number of clusters k and the average dimension l, finding such parameters should be among the major tasks of a clustering algorithm. Moreover, as the illustrative simulations in [10] show, PROCLUS is sensitive to the choice of these input parameters.

2.5 Nonlinear dynamics on invariant subspaces/manifolds. Consequently, for high dimension data sets, projected or subspace clustering seems to be the best solution and, at least conceptually and practically, this has led to significant improvement in the quality of clustering in recent developed algorithms. However, projected clustering seems to be a highly nontrivial task due to the following reasons, in addition to the challenges already faced by clustering even low dimensional data sets:

(C1) Projected clusters can lie in arbitrary subspaces of lower dimensionality;
(C2) Different clusters may form in different subspaces;
(C3) The dimensions of the subspace associated with each projected cluster cannot be identified in advance;
(C4) The number of projected clusters is unknown.

As for (C3), one may attempt to find clusters in all possible subspaces and then to compare the results to obtain an optimal partition of the data set, but this is practically not feasible as the number of all possible subspaces $2^m - 1$ is intractably large for a data set with high dimension m.

The aforementioned additional challenges, in terms of approaching a subspace clustering problem from a nonlinear dynamical system point of view, mean that these subspaces where stable patterns are imbedded are unknown *a-priori*. Finding these subspaces and constructing a nonlinear dynamical system that has such subspaces must be proceeded simultaneously!

We must emphasize here that this concept of subspace clustering is based on the assumption that variables in the data set are independent of each other. If they are nonlinearly correlated, as indeed the case for many data sets, then we will have to search for *submanifold clusters*, here each submanifold cluster consists of a data set D_i and a surface (submanifold) M_i such that points in D_i are close to each other when they are projected onto the surface M_i. Nothing has been achieved so far in this generality.

3 PART: Architecture, models, nonlinear dynamics, simulations

We shall in this section provide a summary of our recent development along the direction of treating subspace clustering as an inverse process of pattern formation of a nonlinear dynamical system to be constructed from a given data set.

3.1 The minimal structure of a nonlinear system for clustering. We now describe our design principle for a nonlinear dynamical system in order to cluster a given data set D in a m dimensional Euclidean space. As already mentioned, there are many nonlinear dynamical systems that share the same dynamical behaviors, and thus we must have a set of design principles that guide us in identifying the minimal structure of such a system. For this purpose, we turn to our central nervous system for inspiration as it does pattern recognition routinely and effectively.

Over the past few decades, there has been significant progress in brain research from a dynamic and adaptive point of view, much of the development is described in the handbook edited by Arbib [7]. In particular, Carpenter and Grossberg have made substantial breakthrough in their design of ART (Adaptive Resonance Theory) neural networks which have been shown to be very effective in self-organized clustering in *full dimensional spaces*. ART was first introduced by Grossberg in 1976 [30], [31] in order to analyze how brain networks can learn in real time about a changing world in a rapid but stable fashion, based on which Capenter and Grossberg [15], [16], [17] developed two classes of ART neural network architectures ART1 and ART2, whose computational performance (dynamics) is described by systems of differential equations. ART1 self-organizes recognition categories for arbitrary sequences of binary input patterns, while ART2 does the same for either binary or continuous inputs. Some other classes of ART neural network architectures such as Fuzzy ART [20], ARTMAP [18], Fuzzy ARTMAP [21] and Gaussian ARTMAP [50] were then developed with increasingly powerful learning and patten recognition capabilities in either an unsupervised or a supervised mode.

Before getting into details of ART, let us now try to answer the question: For a given data set D of n points in the m dimensional Euclidean space, *what the minimal structure a nonlinear dynamical system should have in order to perform effectively and automatically the clustering task* (with respect to all variables, in other words, in the full space):

(MS1) There must be a layer of nodes (neurons) for the purpose of processing inputs (called F_1 layer, also called *input layer or comparison layer*): since each component is expected to play a role in the clustering process, the number of the nodes in this layer should be at least the dimension of the input data;

(MS2) There must be another layer of nodes for the purpose of clustering (called F_2 layer, also called *clustering layer or recognition layer*): each node in this layer should represent a cluster, but the number of clusters is unknown in advance, so we know only that the upper bound of the number of F_2 layer nodes is n;

(MS3) The F_1 layer must be connected to F_2 layer through synaptic connections (*bottom-up weights*), so that the output (activation) of each F_1 node will be weighted appropriately and each F_2 node must then receive a sum of weighted activations from F_1 layer generated by a given input vector;

(MS4) The F_2 layer nodes must be connected together to form a network so that it can automatically select a winner (so that the input vector is assigned to the cluster represented by the winner) from the aforementioned sums of weighted activations from F_1 layer. A natural choice of such a network is a competitive network due to the well-known biological competitive exclusion principle;

(MS5) Once the winner is chosen (and confirmed, see vigilance and rest subsystems discussed below), the corresponding cluster has a new point and the feature (statistical mean) of this cluster must be updated to reflect this learning experience: there must be some mechanism to store the feature of each cluster and this is achieved via the synaptic connections from a F_2 layer node to all nodes of F_1 layer (the so-called *top-down weights*, or *templates*);

(MS6) Both top-down weights and bottom-up weights must be updated following certain learning rules, after each input and clustering process, to learn the experience.

It turns out this is exactly what ART provides, Carpenter and Grossberg designed the F_2 layer as a on-center off-surround network as a special case of a competitive network which follows the winner-take-all paradigm: the node in F_2 with the largest net input (from F_1 layer) becomes the candidate to learn the input pattern. Whether the candidate will learn the input pattern is decided by the vigilance and reset mechanism, which controls the degree of similarity of patterns placed in the same node (cluster).

Figure 3 gives the simplified configuration of an ART architecture. In summary, an ART network includes a choice process and a match process as its key parts. The choice process picks up the most likely category (cluster) for an input pattern. If the template of the chosen category is sufficiently similar to the input pattern to satisfy a predefined vigilance condition, then the category resonates and learns: its template is updated to respond to the new input pattern. Otherwise, the category is reset, and the next most likely category is chosen. If no existing category satisfies the match criterion, then a new category is recruited. Thus, ART incrementally produces categories necessary to represent clusters of input patterns. Note that in ART, the number of clusters does not need to be specified in advance.

Despite the huge success of many applications of ART to practical clustering problems, our simulations (see [10]) show that the current ART architecture has to be modified to perform the task of subspace clustering in high dimensional data sets. In particular, ART finds either 0 or 4 clusters for the data set in Figure 2. This is mainly because ART focuses on similarity of patterns in the **full** dimensional

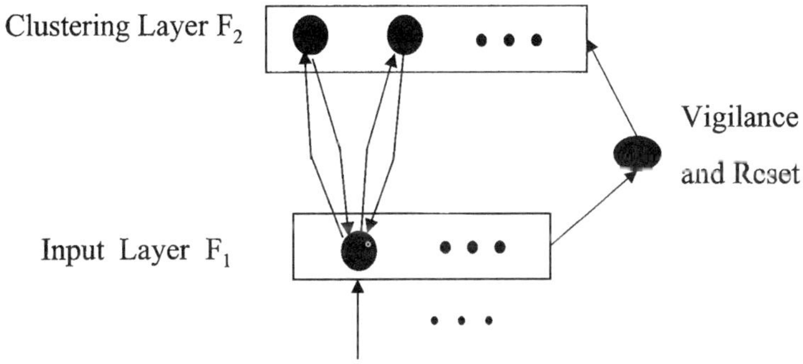

Figure 3 A simplified configuration of ART architecture that consists of an input layer F_1, a clustering layer F_2 and a reset subsystem.

space and a new mechanism must be introduced to deal with the identification of subspaces where clusters are formed.

3.2 PART: Selective output signaling. A candidate of such a new mechanism, essential to the development of PART, called *selective output signaling*, is provided in order to deal with the inherent sparsity in the full space of the data points from many data-mining applications. This selective output signaling mechanism allows the signal generated in a node in the input layer to be transmitted to a node in the clustering layer only when the signal is similar to the top-down weight (template) between the two nodes, and hence PART focuses on dimensions where information can be found.

This also represents the main difference between ART and PART: in ART, the choice process occurs before the match process, while in PART similarity check (between a stored pattern(the template) of a given cluster and an input pattern) takes place before clustering or categorization. This seems to be supported by the most recent neurophysiological finding in [49].

Figure 4 illustrates the basic PART architecture. Similarly to the ART architecture, F_2 layer of PART is a competitive layer which follows the winner-take-all paradigm. The principal difference between PART and ART lies in the functioning of the F_1 layer. In PART, F_1 layer selectively sends signals to nodes in F_2 layer. In other words, a node in F_1 layer can be active relative to some F_2 nodes but inactive relative to other F_2 nodes. To which F_2 node an F_1 node is active is determined by a similarity test between the corresponding top-down weight and the signal generated in the F_1 node. This similarity test plays a key role in the subspace clustering of PART.

Like ART, the vigilance and reset mechanism in PART controls the degree of similarity of patterns in the same cluster, but the similarity measurement in PART is closely related to subspaces involved. In particular, the degree of similarity of patterns is controlled by both vigilance parameter and distance parameter which control the size of dimensions of the projected subspaces and the degree of similarity in a specific dimension involved, respectively. In contrast to PROCLUS, these vigilance and distance parameters are the only required input parameters for the PART algorithm. As our simulations on high dimensional synthetic data in [10]

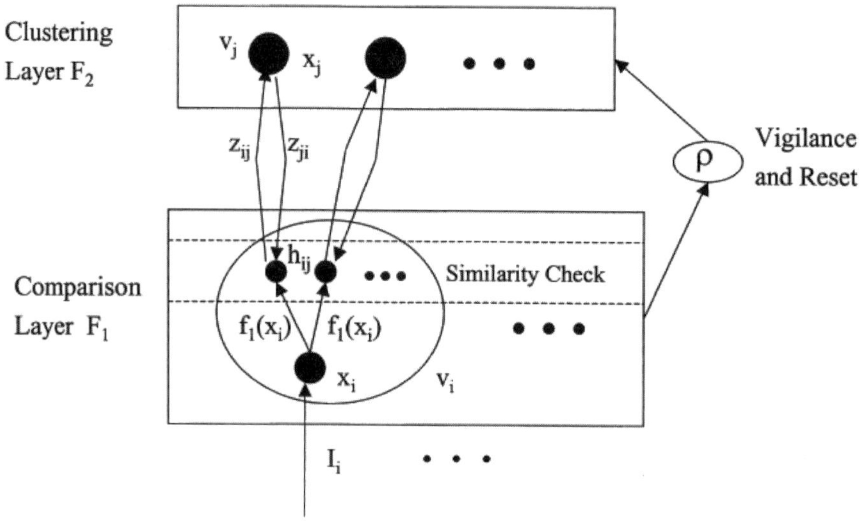

Figure 4 PART architecture. In addition to the usual F_1 layer, F_2 layer and a reset mechanism, there is a hidden layer associated with each F_1 layer node v_i for similarity check to determine whether the node v_i is active relative to an F_2 layer node v_j.

and the recent applications to gene expression and neural spiking trains clustering [47], [36] show, the PART algorithm, with a wide range of input parameters, enables us to fully reproduce the original input clusters.

3.3 PART dynamical model. The PART algorithm developed in [10] is based on the assumptions that the model of PART (a large scale and singularly perturbed system of differential equations coupled with a reset mechanism) has quite regular computational performance described by the following dynamical behaviors, during each learning trial (when a constant input is imposed):

(P1) the winner-take-all paradigm: the F_2 node with the largest bottom-up filter input becomes the winner and only this node is activated after some finite time;

(P2) the selective output signals remain to be constants;

(P3) the synaptic weights are updated following specific formulae;

(P4) the set of dimensions of a specific projected cluster is non-increasing in time.

We now describe the dynamic model in [10]. We denote the nodes in F_1 layer by $v_i, i = 1, \ldots, m$; nodes in F_2 layer by $v_j, j = m+1, \ldots, m+n$; the activation of F_1 node v_i by x_i, the activation of F_2 node v_j by x_j; the bottom-up weight from v_i to v_j by z_{ij}, the top-down weight from v_j to v_i by z_{ji}. The main difference between PART and ART is the introduction of the selective output signal h_{ij} from a node v_i in F_1 to a committed node[1] v_j in F_2 by a similarity check between the top-down weight z_{ji} and the signal $f_1(x_i)$ generated in v_i, where f_1 is a signal function. The introduction of the selective output signaling mechanism in PART allows the signal

[1]We say that a F_2 node is committed if it has learned some input patterns in previous learning traces, and a node is non-committed if it has not learned any input pattern.

generated in a F_1 node to be transmitted to a F_2 node only when the signal is similar to the top-down weight between the two nodes.

The STM (Short Term Memory or activation) equations for nodes in F_1 layer are given by

$$\epsilon_1 \frac{dx_i}{dt} = -x_i + I_i, \quad t > -1, \ 1 \le i \le m, \tag{3.1}$$

where $0 < \epsilon_1 \ll 1$, I_i is the (constant) input imposed on node v_i.

The STM equations for the committed nodes in F_2 layer follow from the on-center off-surround principle in ART and take the form

$$\begin{aligned} \epsilon_2 \frac{dx_j}{dt} &= -x_j + [1 - Ax_j][g(x_j) + T_j] \\ &\quad -[B + Cx_j] \sum_{m+1 \le k \ne j \le m+n} g(x_k), \\ &\quad t \ge 0, m+1 \le j \le m+n, \end{aligned} \tag{3.2}$$

where $0 < \epsilon_2 \ll 1$, g is a signal function to be specified later, A, B, and C are non-negative constants, and the bottom-up filter input T_j is given by

$$T_j = D \sum_{1 \le i \le m} z_{ij} h_{ij} \tag{3.3}$$

with a constant $D > 0$ and the selective output signal h_{ij} to be defined later.

The LTM (Long Term Memory or weight) equations for committed nodes v_j in F_2 layer are

$$\begin{aligned} \delta \frac{dz_{ij}}{dt} &= f_2(x_j)[(1 - z_{ij})Lh_{ij} - z_{ij} \sum_{1 \le k \ne i \le m} h_{kj}], \\ &\quad t \ge 0, \ 1 \le i \le m, m+1 \le j \le m+n, \end{aligned} \tag{3.4}$$

$$\gamma \frac{dz_{ji}}{dt} = f_2(x_j)[-z_{ji}(t) + f_1(x_i)], \quad t \ge 0, \ 1 \le i \le m, m+1 \le j \le m+n, \tag{3.5}$$

where $0 < \delta \ll \gamma = O(1)$, and $L > 0$ is a given constant, f_2 is a signal function.

The LTM equations for non-committed winning node v_j in F_2 layer are

$$\delta \frac{dz_{ij}}{dt} = [1 - z_{ij}]L - z_{ij}(m-1), \quad t \ge 0, \ 1 \le i \le m, \tag{3.6}$$

$$\delta \frac{dz_{ji}}{dt} = -z_{ji} + f_1(x_i), \quad t \ge 0, \ 1 \le i \le m. \tag{3.7}$$

The term h_{ij} is defined as follows:

$$h_{ij} = h_\sigma(d(f_1(x_i), z_{ji}))l_\zeta(z_{ij}), \quad t \ge 0, \ 1 \le i \le m, m+1 \le j \le m+n, \tag{3.8}$$

where we take

$$d(a, b) = |a - b| \quad \text{for any } a, b \in \mathbb{R}, \tag{3.9}$$

and for a given constant c, h_c is given by

$$h_c(\xi) = \begin{cases} 1 & \text{if } \xi \le c, \\ 0 & \text{if } \xi > c, \end{cases} \tag{3.10}$$

and l_c is given by

$$l_c(\xi) = 1 - h_c(\xi). \tag{3.11}$$

In the definition of Eq. (3.8), we require the *distance parameter* $\sigma > 0$ and

$$0 \le \zeta < L/(L - 1 + m). \tag{3.12}$$

In the PART architecture, the above dynamical process is coupled with a reset mechanism: A winning F_2 node will be reset so that it will always be inactive during the remainder of the current trial if it does not satisfy some vigilance conditions. In particular, a winning (active) F_2 node v_j will be reset if at any given time $t \geq 0$, the degree of match r_j defined by

$$r_j = \sum_i h_{ij} \tag{3.13}$$

is less than a prescribed vigilance. Namely, reset occurs if and only if

$$r_j < \rho,$$

here $\rho \in \{1, 2, \ldots, m\}$ is a *vigilance parameter.*

In what follows, we are going to make the following set of standing assumptions:

(H1). The constants A, B and C satisfy

$$A = 1, B = 0, C > 0. \tag{3.14}$$

(H2). The signal function $f_1 : R \to R$ is nondecreasing and satisfies the Lipschitz condition (with a given constant $M > 0$)

$$|f_1(x) - f_1(y)| \leq M|x - y|, \quad x, y \in \mathbb{R}. \tag{3.15}$$

(H3). The signal functions f_2 and g satisfy

$$f_2(x) = \begin{cases} 1 & \text{if } x \geq \eta, \\ 0 & \text{if } x < \eta, \end{cases} \tag{3.16}$$

$$g(x) = \begin{cases} 1 & \text{if } x \geq \theta, \\ 0 & \text{if } x < \theta, \end{cases} \tag{3.17}$$

where θ and η are two thresholds. Usually we require $\theta \leq \eta$.

Assumption (H1) is made for the sake of simplifying the presentation, and is in fact a rescaling procedure so that the maximum level of activation of each node in F_2 layer is 1. A and C can be any positive constants, and B can be any non-negative constant, as long as the ranges of signal functions and thresholds involved are modified accordingly. (H2) is an assumption satisfied by all signal functions used in the literature. (H3) requires all signal functions to be step functions with thresholds, and this reflects the on-or-off characteristic of neurons, and these are required due to some technical issues in the mathematical proofs.

Define

$$T^{max} = \frac{DmL}{L - 1 + m}, \quad T^{min} = \frac{DL}{L - 1 + m}. \tag{3.18}$$

We then let T_{max} and T_{min} be two arbitrarily given constants satisfying

$$T_{max} \geq T^{max}, \tag{3.19}$$

$$T_{min} \leq T^{min}. \tag{3.20}$$

In other words, T^{max} and T^{min} are respectively the upper bound and lower bound of all possible nonzero bottom-up filter inputs T_j under perfect learning (see definition below).

3.4 Transient and long-term dynamics. The following technical Lemma plays an important role in describing the dynamical behaviors of the network.

Lemma 3.1 *Consider the system*

$$\begin{aligned} \epsilon_2 \frac{dx_j}{dt} &= -x_j + [1 - Ax_j][g(x_j) + T_j] \\ &\quad -[B + Cx_j] \sum_{m+1 \le k \ne j \le m+n} g(x_k), \\ &\quad m+1 \le j \le m+n, \end{aligned} \tag{3.21}$$

where $\epsilon_2 > 0$ and all T_j are constants, $(A,\ B,\ C)$ and g satisfy conditions (3.14)-(3.17). Assume that there exists $m+1 \le J \le m+n$ such that $T_J/(1+T_J+C) \le \theta$, $0 \le T_j < T_J$ and $0 \le x_j(0) \le x_J(0) < \theta$ for all $m+1 \le j \ne J \le m+n$. Then we have the following conclusions:

(a) If $\frac{T_J}{1+T_J} > \theta$, then we have the following results:

(i) *For all $j \ne J$, $x_j(t) < \theta$ when $t \ge 0$, and*

$$\lim_{t\to\infty} x_j(t) = \frac{T_j}{1+T_j+C};$$

(ii) *$x_J(t)$ is increasing for $t \ge 0$, and*

$$\lim_{t\to\infty} x_J(t) = \frac{1+T_J}{2+T_J};$$

(iii) *For any given $\eta \in [\theta, \frac{1+T_J}{2+T_J})$, there exists $T > 0$ such that $x_J(t) < \eta$ when $t < T$ and $x_J(t) \ge \eta$ when $t \ge T$. Moreover, we can make $T < 1$ if ϵ_2 is sufficiently small;*

(b) If $\frac{T_J}{1+T_J} \le \theta$, then for all $m+1 \le j \le m+n$ and $t \ge 0$, $x_j(t) < \theta$.

Note that an F_2 node v_j is active at time t if and only if $f_2(x_j(t)) > 0$. For every $j \ne J$, $x_j(t) < \theta \le \eta$ implies that $f_2(x_j(t)) = 0$. Therefore, Lemma 3.1-(a)-(i) implies that all F_2 nodes v_j with $j \ne J$ are always inactive. Using the same argument, Lemma 3.1-(a)-(iii) implies that F_2 node v_J is activated after some finite time T. This is the winner-take-all paradigm.

Lemma 3.1 shows the winner-take-all paradigm in a simple case where all bottom-up filter inputs (T_j's) are fixed constants. For the full PART model, of course, T_j vary in time, and we have the following

Theorem 3.2 *Consider the system of differential equations (3.1)-(3.14) and assume that $0 < \epsilon_1, \epsilon_2, \delta \ll 1$, $\gamma = O(1)$, and conditions (H1), (H2) and (H3) hold with θ and η satisfying*

$$\frac{T_{max}}{1+T_{max}+C} < \theta < \frac{T_{min}}{1+T_{min}}, \tag{3.22}$$

$$\theta \le \eta < \frac{1+T_{min}}{2+T_{min}}. \tag{3.23}$$

Assume also that there exists $m+1 \le J \le m+n$ such that $0 \le T_j(0) < T_J(0)$ and $0 \le x_j(0) \le x_J(0) < \theta$ for all $m+1 \le j \ne J \le m+n$, and $T_{min} \le T_J(0) \le T_{max}$. Then when ϵ_1, ϵ_2 and δ are sufficiently small, the following results hold:

(P1) *(Inhibition) For $j \ne J$ and $t \ge 0$, $x_j(t) < \theta$ and $f_2(x_j(t)) = 0$;*

(P2) *(Excitation) There exists T with $0 < T < 1$ such that $x_J(t) < \eta$ and $f_2(x_J(t)) = 0$ when $t < T$, and $x_J(t) \geq \eta$ and $f_2(x_J(t)) = 1$ when $t \geq T$;*

(P3) *(Similarity Invariance) For any $t \geq 0$,*

$$h_{ij}(t) = h_{ij}(0); \tag{3.24}$$

(P4) *(Long-term and Transient Learning) For any i and $j \neq J$, $z_{ij}(t)$ and $z_{ji}(t)$ remain unchanged for all $t \geq 0$. But*

$$\lim_{t\to\infty} z_{iJ}(t) = \begin{cases} 0 & \text{if } h_{iJ}(0) = 0, \\ \frac{L}{L-1+|X|} & \text{if } h_{iJ}(0) = 1, \end{cases} \tag{3.25}$$

where

$$X = \{i \in \Lambda_1 : h_{iJ}(0) = 1\}, \tag{3.26}$$

and $|X|$ denotes the number of elements of the set X, and

$$\lim_{t\to\infty} z_{Ji}(t) = f_1(I_i). \tag{3.27}$$

Moreover, writing $z_{ij}^{\epsilon_1,\epsilon_2,\delta}$ and $z_{ji}^{\epsilon_1,\epsilon_2,\delta}$ to indicate explicitly the dependence on $(\epsilon_1, \epsilon_2, \delta)$, we have

$$\lim_{\delta\to 0} z_{iJ}^{\epsilon_1,\epsilon_2,\delta}(1) = \begin{cases} 0 & \text{if } h_{iJ}(0) = 0, \\ \frac{L}{L-1+|X|} & \text{if } h_{iJ}(0) = 1, \end{cases} \tag{3.28}$$

$$\lim_{\epsilon_1\to 0} z_{Ji}^{\epsilon_1,\epsilon_2,\delta}(1) = (1-\alpha)z_{Ji}(0) + \alpha f_1(I_i), \tag{3.29}$$

where $\alpha = 1 - e^{-1/\gamma}$;

(P5) *(Subspace Identification) Define $D_j(t) = \{i : l_\zeta(z_{ij}(t)) = 1\}$. Then*
- (P5a) *$D_j(t) = D_j(0)$ for any $j \neq J$;*
- (P5b) *$D_J(t_2) \subseteq D_J(t_1)$ if $t_2 \geq t_1 \geq 0$;*
- (P5c) *$D_J(t) = D_J(1) = X$ for all $t \geq 1$.*

Theorem 3.2 describes the dynamics of PART during a trial. (P1) and (P2) in Theorem 3.2 verify the winner-take-all paradigm. Recall that an F_2 node v_j is active at time t if and only if $f_2(x_j(t)) > 0$. (P1) implies all F_2 nodes v_j with $j \neq J$ are always inactive (unless a reset process occurs). (P2) implies that F_2 node v_J is activated after some finite time T. In other words, the F_2 node (v_J) with the largest bottom-up filter input (T_J) becomes the winner after some finite time T.

(P3) implies all the selective output signals remain to be constants. It follows from this property that the degree of match $r_j(t)$ $(= \sum_i h_{ij}(t))$ keeps constant. This suggests an alternative implementation of PART net: reset in parallel all non-matching F_2 nodes at $t = 0$, then select the winner from remaining F_2 nodes so that the winner must satisfy the match criterion. This will speed up the network computation since the serial search process (caused by repeated resets to successive winners in some situations, in particular, when a new pattern is input) is omitted.

Eqs. (3.28) and (3.29) (in (P4)) are the learning formulae used in PART algorithm, with $\alpha = 1 - e^{-1/\gamma}$ being the learning rate.

Recall that $D_j(t)$ is the set of dimensions of projected subspace associated with cluster C_j (represented by F_2 node v_j) at time t. (P5) shows that the set of dimensions is non-increasing during the learning. In particular, the resulting dimensions D_J of winner v_J after the learning trial corresponds to those F_1 nodes which are active relative to the winner v_J. This non-increasing property of dimensions contributes to stabilizing learning in response to arbitrary sequences of input patterns.

A necessary and sufficient condition for an F_2 node v_j to be activated is given in [13], as described below:

Lemma 3.3 *Assume that* $0 < \epsilon_1, \epsilon_2, \delta \ll 1$, $\gamma = O(1)$, *and conditions (H1), (H2) and (H3) hold with* θ *and* η *satisfying*

$$\frac{T^{max}}{1 + T^{max} + C} < \theta < \frac{T^{min}}{1 + T^{min}}, \tag{3.30}$$

$$\theta \leq \eta < \frac{1}{2 - \theta}. \tag{3.31}$$

Assume also that there exists $m + 1 \leq J \leq m + n$ *such that* $0 \leq T_j(0) < T_J(0)$ *and* $0 \leq x_j(0) \leq x_J(0) < \theta$ *for all* $j \neq J$, *and* $T_J(0) \leq T^{max}$. *Then when* ϵ_1, ϵ_2 *and* δ *are sufficiently small, we have the following results:*

(NS1) *If* $T_J(0) > \frac{\theta}{1-\theta}$, *then all conclusions of Theorem 3.2 hold;*

(NS2) *If* $T_J(0) \leq \frac{\theta}{1-\theta}$, *then for all* $m + 1 \leq j \leq m + n$ *and* $t \geq 0$, $x_j(t) < \theta$ *and* $f_2(x_j(t)) = 0$.

Note that the condition $T_J(0) \leq \frac{\theta}{1-\theta}$ under which no F_2 node will be activated (and therefore the winner-take-all paradigm fails) can hardly occur: the second inequality of Eq. (3.30) and condition $T_J(0) \leq \frac{\theta}{1-\theta}$ imply that $T_J(0) \leq \frac{\theta}{1-\theta} < T^{min}$, but $T_J(0) < T^{min}$ is impossible under perfect learning, described below.

Theorem 3.2 requires that there exists $m + 1 \leq J \leq m + n$ such that $T_{min} \leq T_J(0) \leq T_{max}$ and $0 \leq T_j(0) < T_J(0)$. It is natural to ask whether these assumptions can be satisfied in a sequence of arbitrary trials. To answer this problem, we say that a winning F_2 node v_J has a *perfect learning* at a trial if either v_J is a committed node and, at the end of the trial z_{iJ}, satisfies

$$z_{iJ} = \begin{cases} 0 & \text{if } h_{iJ}(0) = 0, \\ \frac{L}{L-1+|X|} & \text{if } h_{iJ}(0) = 1, \end{cases} \tag{3.32}$$

or v_J is a noncommitted node and, at the end of the trial z_{iJ}, satisfies

$$z_{iJ} = \frac{L}{L - 1 + m}, \tag{3.33}$$

for all $1 \leq i \leq m$. It was shown in [13] that if an F_2 node v_j has perfect learning in last trial, then at the beginning of the current trial, either

$$T_{min} \leq T^{min} \leq T_j(0) \leq T^{max} \leq T_{max}$$

or

$$T_j(0) = 0$$

for any input pattern.

The assumption that there exists J such that $0 \leq T_j(0) < T_J(0)$ for all $m+1 \leq j \neq J \leq m + n$ is not always satisfied. When this assumption is not satisfied it is possible that more than one committed F_2 nodes are activated or no committed F_2 node is activated. In the case where no committed F_2 node is activated, a noncommitted F_2 node will be selected as the winner. In the case where more than one committed F_2 nodes are activated, one usually has to increase the vigilance gradually. At each given vigilance level, some F_2 nodes with lower degree of match are reset until either a unique F_2 node with the largest degree of match becomes the winner (when there is a unique active F_2 node with the largest degree of match) or all active F_2 nodes are reset and then a noncommitted F_2 node is selected as the winner (when there are two or more active F_2 nodes with the same largest degree of match). In [13], a sufficient condition to ensure no more than one F_2 node can be activated at the same time. This issue will also be addressed in PART-D.

The proof of Theorem 3.2 is technical and long. Here are the main ideas and steps. First of all, we shall show that during a particular trial, the similarity measure (given by h_σ) between the output of an F_1 node and the corresponding component of the template of a target F_2 node remains to be a constant (persistence), this is the essential part in order to maintain the stability of the learning and categorizing process for each input. With this in mind, then the description of the dynamics in the F_2 layer (that decides the candidate for a given input) becomes relatively easy due to the "competitive exclusion principle": the total selective output signals from all F_1 nodes to all F_2 nodes have been decided already and thus the competitive nature of the F_2 layer selects the candidate based on the winner-take-all paradigm. Once this is shown, the equations for the updating of the bottom-up and top-down weights become simple and decoupled ordinary differential equations that can be explicitly solved.

What makes the proof technically difficult is that we have to prove the "similarity measure persistence" and the "competitive exclusion principle" in two steps: we first show the "similarity measure persistence" holds true until F_2 layer selects its winner, and then we have to show simultaneously that the winner remains to be a winner and the similarity measures remain to be constants, and we do so by way of contradiction: violation of one of the above statements at some point will lead to a contradiction.

3.5 PART algorithm and examples. We now summarize the PART algorithm based on the transient dynamics of the PART dynamical model. The algorithm was designed in [12] and was later refined in [13], from which the current form is taken.

A1. Input Processing and Select Output Signals from Input Layer: For simplicity, through this subsection, we use the identical function $f_1(x_i) = x_i$, then we have at an equilibrium $x_i = I_i$, $f_1(x_i) = I_i$, where $I = (I_1, \cdots, I_m)$ is an input pattern. Therefore, with σ being a distance vigilance parameter and ζ a small threshold usually taken as 0, we have

$$h_{ij} = \begin{cases} 1 & \text{if } |f_1(x_i) - z_{ji}| \leq \sigma \text{ and } z_{ij} > \zeta, \\ 0 & \text{otherwise.} \end{cases} \tag{3.34}$$

A2. Activation, Inhibition, and Identification of a Potential Cluster: We compute the bottom-up filter input T_j to the committed F_2 node v_j (assuming $D = 1$) and then select the winner according to the rule as follows:

Let $\Gamma = \{T_k : F_2$ node v_k is committed and has not been reset on the current trial$\}$, then node v_J is a winner either if $\Gamma \neq \emptyset$ and $T_J = \max \Gamma$, or if $\Gamma = \emptyset$ and node v_J is the next noncommitted node in F_2 layer.

A3. Confirmation, Vigilance and Reset: A winning F_2 node v_J is reset if the matching degree r_J is less than a vigilance parameter ρ. Otherwise, the winner passes the vigilance test and is ready to learn the input pattern.

A4. Learning: For the winner v_J which has passed the vigilance test, update its bottom-up weights z_{iJ} and its top-down weights z_{Ji} as follows:

If v_J is a committed winning F_2 node, then

$$z_{iJ}^{new} = \begin{cases} L/(L-1+|X|) & \text{if } h_{iJ} = 1, \\ 0 & \text{if } h_{iJ} = 0, \end{cases} \tag{3.35}$$

$$z_{Ji}^{new} = (1-\alpha)z_{Ji}^{old} + \alpha I_i, \tag{3.36}$$

where $0 \le \alpha \le 1$ is the learning rate.

If v_J is a noncommitted F_2 node, then

$$z_{iJ}^{new} = L/(L-1+m), \tag{3.37}$$

$$z_{Ji}^{new} = I_i. \tag{3.38}$$

A5. Identification of Subspaces: Each committed F_2 node v_j represents a projected cluster C_j. The set D_j of the associated dimensions of the projected cluster C_j is determined by $l_\varsigma(z_{ij})$ according to the following formula:

$$\text{The dimension } i \in D_j \text{ if and only if } l_\varsigma(z_{ij}) = 1. \tag{3.39}$$

A6. Outliers: Theoretically, F_2 layer can have arbitrarily many nodes in the PART architecture, therefore it can categorize arbitrarily many input patterns. But due to the resource restriction, the number of nodes in F_2 layer has to be limited. On the other hand, in many clustering problems, there are always some data points (called outliers) which do not cluster well. Therefore, we add a special node in PART module, called outlier node. We simply put into the outlier node all data points that cannot be clustered into F_2 nodes.

The step-by-step PART algorithm is presented in detail as follows.

PART Algorithm:

0. Initialization:
   Take the number m of nodes in F_1 layer as the number of dimensions of input data.
   Choose the number n of nodes in F_2 layer much larger than the expected number of clusters.
   Set the internal parameters L, α and ς and maximum iteration M.
   Choose the external input parameters ρ and σ.
1. Set all F_2 nodes as being noncommitted.
2. For each data point in input data set, do Steps 2.1-2.6.
   2.1. Compute h_{ij} for all F_1 nodes v_i and committed F_2 nodes v_j. If all F_2 nodes are noncommitted, go to Step 2.3.
   2.2. Compute T_j for all committed F_2 nodes v_j.
   2.3. Select the winning F_2 node v_J. If no F_2 node can be selected, put the data point into outlier and then continue to do Step 2.
   2.4. If the winner is a committed node, compute r_J, otherwise go to Step 2.6.
   2.5. If $r_J \ge \rho$, go to Step 2.6, otherwise reset the winner v_J and go back to Step 2.3.
   2.6. Set the winner v_J as the committed, and update the bottom-up and top-down weights for winner node v_J.
3. Repeat Step 2 M times.
4. For each cluster C_j in F_2 layer, compute the associated dimension set D_j.

The time cost of PART algorithm is $O(mnNM)$, where m is the number of dimensions of data space, n is the number of F_2 nodes, N is the number of all data points and M is the number of iterations. Our simulations on high dimensional synthetic data showed that the categories in F_2 layer became stable after only a few iterations.

Table 1 gives the permissible range and the suggested sample values of parameters in the above PART algorithm. For large data sets, some random data points

Table 1 List of parameters for the PART algorithm.

PARAMETER	PERMISSIBLE RANGE	SAMPLE VALUE
L	$L > 1$	2
α	$0 \leq \alpha \leq 1$	0.1
ζ	$0 \leq \zeta < L/(L-1+m)$	0
ρ	$1 \leq \rho \leq m$	NA
σ	$\sigma \geq 0$	NA

may be correlated in several dimensions. However, it is very unlikely that a large number of random data points are correlated in a large set of dimensions. Therefore, we should choose ρ large enough to eliminate the randomness, but smaller than or equal to the number of dimensions of any possible projected cluster. As we do not know in advance the numbers of dimensions of projected clusters, we can choose ρ as small as possible, but large enough to eliminate the randomness.

We now provide a small-scale example to illustrate the PART algorithm.

Example 3.4 Consider the data set {(1, 2, 3, 4), (2, 3, 4, 5), (3, 4, 5, 6), (4, 5, 6, 7), (6, 7, 8, 9), (1, 2, 3, 4), (3, 4, 5, 6), (2, 3, 4, 5), (6, 4, 5, 6), (4, 2, 3, 1), (6, 7, 1, 2), (4, 5, 6, 7)} of 12 points in 4-dimensional space. Note that a few points appear twice. We apply PART algorithm with $\rho = 1$, $\sigma = 0.5$, $L = 2$, $\zeta = 0$ and $\alpha = 0.1$. The choice of $\rho = 1$ requires two points must be close to each other with respect to at least one dimension, and the choice of $\sigma = 0.5$ is made since all numbers involved in the data set are integers. We set $m = 4$ (the dimension of data space) and $n = 12$ (the maximal number of clusters).

Step 1. $I = (1,2,3,4)$. Since no F_2 node is committed yet, it is natural to select v_5 (the first node in F_2 layer) to learn the input pattern. We have bottom-up weights $z^{new}_{5(bu)} = (z^{new}_{i5})^4_{i=1}$ with

$$z^{new}_{i5} = \frac{2}{2-1+4} = \frac{2}{5},$$

and the top-down weights

$$z^{new}_{5(td)} = (z^{new}_{5i})^4_{i=1} = (1,2,3,4).$$

Step 2. $I = (2,3,4,5)$. In this case,

$$h_5 = (h_{15}, h_{25}, h_{35}, h_{45}) = (0,0,0,0)$$

since $|I_i - z_{5i}| = 1 > 0.5 = \sigma, \quad i = 1, \cdots, 4$. We have $r_5 = 0 < 1 = \rho$ and hence we select the next noncommitted F_2 node v_6 to learn the pattern. We have

$$z^{new}_{6(bu)} = (z^{new}_{i6})^4_{i=1} = (\frac{2}{5}, \frac{2}{5}, \frac{2}{5}, \frac{2}{5}),$$

$$z^{new}_{6(td)} = (z^{new}_{6i})^4_{i=1} = (2,3,4,5).$$

Similarly, we have

Step 3. $I = (3,4,5,6)$. v_7 is the noncommitted F_2 node to learn the pattern,

$$z^{new}_{7(bu)} = (\frac{2}{5}, \frac{2}{5}, \frac{2}{5}, \frac{2}{5}),$$

$$z_{7(td)}^{new} = (3,4,5,6).$$

Step 4. $I = (4,5,6,7)$. v_8 is the noncommitted F_2 node to learn the pattern,

$$z_{8(bu)}^{new} = (\frac{2}{5}, \frac{2}{5}, \frac{2}{5}, \frac{2}{5}),$$
$$z_{8(td)}^{new} = (4,5,6,7).$$

Step 5. $I = (6,7,8,9)$. v_9 is the noncommitted F_2 node to learn the pattern,

$$z_{9(bu)}^{new} = (\frac{2}{5}, \frac{2}{5}, \frac{2}{5}, \frac{2}{5}),$$
$$z_{9(td)}^{new} = (6,7,8,9).$$

Step 6. $I = (1,2,3,4)$. We have
$h_5 = (1,1,1,1)$ since $|I_i - z_{5i}| = 0 < \sigma$ and $z_{i5} = 2/5 > 0$,
$T_5 = \sum_{i=1}^{4} z_{i5}h_{i5} = 8/5$,
$h_6 = h_7 = h_8 = h_9 = (0,0,0,0)$ since $|I_i - z_{ji}| = 1 > \sigma$ for $i = 1,2,3,4,\quad j = 6,7,8,9$,
$T_6 = T_7 = T_8 = T_9 < T_5$.
So v_5 is the winning F_2 node. Since $r_5 = 4 \geq 1 = \rho$ we conclude that v_5 learns the input pattern

$$z_{5(bu)}^{new} = (\frac{2}{5}, \frac{2}{5}, \frac{2}{5}, \frac{2}{5}),$$
$$z_{5(td)}^{new} = (1-\alpha)z_{5(td)}^{old} + \alpha I = 0.9(1,2,3,4) + 0.1(1,2,3,4) = (1,2,3,4).$$

Similarly, we have
Step 7. $I = (3,4,5,6)$. v_7 is the winning F_2 node and learns the input pattern,

$$z_{7(bu)}^{new} = (\frac{2}{5}, \frac{2}{5}, \frac{2}{5}, \frac{2}{5}),$$
$$z_{7(td)}^{new} = (3,4,5,6).$$

Step 8. $I = (2,3,4,5)$. v_6 is the winning F_2 node and learns the input pattern,

$$z_{6(bu)}^{new} = (\frac{2}{5}, \frac{2}{5}, \frac{2}{5}, \frac{2}{5}),$$
$$z_{6(td)}^{new} = (2,3,4,5).$$

Step 9. $I = (6,4,5,6)$. We have
$h_5 = h_6 = h_8 = (0,0,0,0)$,
$h_7 = (0,1,1,1)$,
$h_9 = (1,0,0,0)$,
$T_7 = \sum_{i=1}^{4} z_{i7}h_{i7} = 6/5$ (using $z_{i7} = (z_{7(bu)}^{new})_i$ in Step 3),
$T_9 = \sum_{i=1}^{4} z_{i9}h_{i9} = 2/5$ (using $z_{i9} = (z_{9(bu)}^{new})_i$ in Step 5).
$T_9 < T_7$, so v_7 is the winning F_2 node.
$r_7 = 3 \geq 1$, so v_7 learns the input pattern,

$$z_{7(bu)}^{new} = (0, \frac{2}{4}, \frac{2}{4}, \frac{2}{4}) = (0, \frac{1}{2}, \frac{1}{2}, \frac{1}{2}),$$
$$z_{5(td)}^{new} = (1-\alpha)z_{5(td)}^{old} + \alpha I = 0.9(3,4,5,6) + 0.1(6,4,5,6) = (3.3,4,5,6).$$

Step 10. $I = (4,2,3,1)$. We have
$h_5 = (0,1,1,0)$,
$h_6 = h_7 = h_9 = (0,0,0,0)$,
$h_8 = (1,0,0,0)$,

Table 2 Dimensions and numbers of data points of input clusters for a data set in a 100-dimensional space.

Input	Dimensions	Points
1	1, 6, 10, 12, 15, 17, 31, 36, 37, 38, 45, 46, 52, 54, 58, 61, 67, 79, 81, 99	5250
2	1, 3, 5, 6, 8, 10, 11, 13, 15, 16, 17, 18, 19, 20, 27, 28, 30, 38, 59, 68, 70, 84, 85, 92	1944
3	3, 5, 10, 11, 12, 15, 16, 18, 23, 37, 43, 52, 58, 61, 74, 81, 99	4652
4	2, 3, 6, 9, 12, 13, 15, 31, 43, 51, 70, 91, 95	3006
5	1, 2, 3, 8, 9, 11, 12, 13, 16, 18, 28, 41, 48, 51, 74, 82	2676
6	1, 2, 3, 5, 6, 7, 8, 9, 10, 11, 12, 13, 14, 15, 16, 20, 30, 40, 51, 54, 57, 58, 74, 77, 84, 90, 99, 100	1472
Outliers	-	1000

$T_5 = \sum_{i=1}^4 z_{i5}h_{i5} = 4/5$ (using $z_{i5} = (z_{5(bu)}^{new})_i$ from Step 6),
$T_8 = \sum_{i=1}^4 z_{i8}h_{i8} = 2/5$ (using $z_{i8} = (z_{8(bu)}^{new})_i$ from Step 4).
$T_8 < T_5$, so v_5 is the winning F_2 node.
$r_5 = 2 \geq 1$, so v_5 learns the input pattern,

$$z_{5(bu)}^{new} = (0, \frac{2}{3}, \frac{2}{3}, 0),$$

$$z_{5(td)}^{new} = (1-\alpha)z_{5(td)}^{old} + \alpha I = 0.9(1,2,3,4) + 0.1(4,2,3,1) = (1.3, 2, 3, 3.7).$$

Similarly, we have v_9 learns the input pattern $(6,7,1,2)$ and v_8 learns the input pattern $(4,5,6,7)$, and we obtain five projected clusters

$$\begin{aligned} C_5 &= \{(4,2,3,1),(1,2,3,4),(1,2,3,4)\}, \\ C_6 &= \{(2,3,4,5),(2,3,4,5)\}, \\ C_7 &= \{(6,4,5,6),(3,4,5,6),(3,4,5,6)\}, \\ C_8 &= \{(4,5,6,7),(4,5,6,7)\}, \\ C_9 &= \{(6,7,8,9),(6,7,1,2)\}, \end{aligned}$$

the corresponding subspaces are $(0,1,1,0)$, $(1,1,1,1)$, $(0,1,1,1)$, $(1,1,1,1)$ and $(1,1,0,0)$, where 1 and 0 indicate that the corresponding dimension is used or not used respectively.

3.6 Simulations and comparisons. We now provide a simulation example on high dimensional synthetic data. We refer to [10] for more simulation examples and performance comparisons with other algorithms.

The high dimensional synthetic data are generated via the method introduced by Aggarwal et al in 1999 [2]. In the experiment reported below, we use an input data file with 20,000 data points in a 100-dimensional space, which has 6 clusters generated in 20, 24, 17, 13, 16 and 28-dimensional subspaces respectively. Table 2 shows the associated dimensions (the subspace where cluster is formed) and the number of data points of each cluster in the input file. The data points are presented in random order in the experiment. We report the results, with emphasis on four different aspects: number of clusters found, dimensions found, centers of clusters found, and the contingency table [34] of input clusters (original clusters) and output clusters (clusters found). Table 3 an Table 4 show the simulation results with $\rho = 10$. Note that in the reported results we have treated as outliers the data

Table 3 Dimensions and numbers of the data points of output clusters.

Output	Dimensions	Points
1	10, 12, 17, 37, 46, 58, 61, 79, 81, 99	5144
2	5, 8, 13, 15, 16, 18, 30, 70, 85, 92	1878
3	3, 5, 10, 12, 15, 23, 43, 58, 74, 99	4412
4	2, 6, 9, 12, 15, 31, 43, 70, 91, 95	2716
5	1, 2, 3, 9, 13, 16, 18, 28, 41, 74	2608
6	1, 3, 5, 9, 15, 51, 57, 58, 84, 90	1186
Outliers	-	2056

Table 4 Contingency table of input clusters and output clusters. Entry (i, j) denotes the number of data points that are common to output cluster i and input cluster j. The table shows that the original input cluster structure is successfully identified.

Output\Input	1	2	3	4	5	6	Outliers	Sums
1	5144	0	0	0	0	0	0	5144
2	0	1878	0	0	0	0	0	1878
3	0	0	4412	0	0	0	0	4412
4	0	0	0	2716	0	0	0	2716
5	0	0	0	0	2608	0	0	2608
6	0	0	1	0	0	1185	0	1186
Outliers	106	66	239	290	68	287	1000	2056
Sums	5250	1944	4652	3006	2676	1472	1000	20000

points in the F_2 categories with very small sizes (less than 1.2% of total data points in this experiment). The simulation results show that the PART algorithm succeeds in finding the exact number of original clusters and in finding almost exact centers of all original clusters. The dimensions found are not identical to those of the original clusters, but these found dimensions are contained as subsets of the associated dimensions of original clusters. These subsets are sufficiently large so that, after a further reassignment procedure, we are able to reproduce the original clusters from the found cluster centers, the found number of clusters and the found dimensions. Table 5 shows the reassignment results. The reassignment procedure reassigns every data point to its closest cluster center according to the Manhattan segmental distance relative to the found dimensions.[2]

4 How and why PART works: Delay adaptation and PART-D

As discussed, ART consists of a selection process and a similarity check process to reinforce a learning mechanism: the selection process picks up the most likely category (cluster representation) for an input pattern, and a verification process checks the similarity between the chosen category's template and input pattern before updating the template in response to the new input pattern.

[2]Here, the Manhattan segmental distance is defined as follows: For any two points $x_1 = (x_{11}, ..., x_{1d})$, $x_2 = (x_{21}, ..., x_{2d})$, and a set of dimensions D, $|D| \leq d$, the Manhattan segmental distance between x_1 and x_2 relative to D is $d_D(x_1, x_2) = \sum_{i \in D} |x_{1i} - x_{2i}|/|D|$.

Table 5 Reassignment result according to the found cluster centers and dimensions. The result shows that the original input clusters are fully reproduced.

Output\Input	1	2	3	4	5	6	Outliers	Sums
1	5250	0	0	0	0	0	234	5484
2	0	1944	0	0	0	0	194	2138
3	0	0	4651	0	0	0	125	4776
4	0	0	0	3006	0	0	148	3154
5	0	0	0	0	2676	0	123	2799
6	0	0	1	0	0	1472	176	1649
Sums	5250	1944	4652	3006	2676	1472	1000	20000

PART provides an additional capacity of identifying clusters and patterns in subspaces by introducing the new selective signal output mechanism. This mechanism requires checking the similarities (between the templates and a new input pattern) before deciding which dimensions will be used for the purpose of clustering (selecting a category node). This reversing the order of similarity check and pattern recognition seems to be supported by the recent neurophysiological findings in [42], although the identity of the neurophysiological mechanism that determines the similarity between a sensory input and a stored memory precept is not presently known.

In [41], a more general version of PART, called *PART-D*, is developed primarily in order to provide a neuro-physiological basis for the determination of similarity necessary for the match-based learning algorithm. It shows that the similarity criteria can be closely linked to the concept that conduction delays in the nervous system can be adapted.

This modified PART architecture is based on the observation that "The site of learning in most neural networks models occurs in either synaptic weights or synaptic delays or both independently. However, the physiology of synapses and dendrites in many systems indicates that these quantities should co-vary" [41]. This observation leads to a novel mathematical model for subspace clustering: a new class of large-scale delay differential equations with state-dependent delays.

4.1 PART-D: System of delay differential equations with adaptive delay. The crucial question in PART architecture is how to calculate the total amount of inputs from all of the neurons in the F_1 layer to a given cluster neuron. This is essential for identifying a winner candidate cluster node that represents the input vector and is consequently closely related to the nature of the clustering criterion to be discovered by the network. How to deal with this calculation distinguishes PART from its ancestor ART, and PART-D is designed to offer a neurophysiological explanation why such a distinction is important for subspace clustering.

In PART-D, we use the properties of synapses and signal conduction in a passive dendritic to refine the calculation of the total input to a given cluster neuron from all input processing neurons by taking into account the effects of transmission time delays and their impacts on the signal loss.

Since neurons in F_1 and F_2 are physically separated, the effects of conduction time delays must be taken into account. Even for passive dendrites, background synaptic activity and voltage–dependent nonlinearities ensure that apical dendrites function as a dynamic delay line [4]. Unsupervised, activity–dependent changes in time delays, i.e. delay adaptation, are known to occur in the nervous system [46, 45]. Thus it is possible that learning occurs not only through changes in the synaptic weights, but also because of delay adaptation. In other words learning can be associated with either morphological changes in neurons and/or changes in synapses ([23], [24], [25], [40], [51]).

We make two assumptions about the role of delay adaptation for data clustering:

(AD1) The evolution of the delay is driven by a dissimilarity measure that is related to the status of the input neurons in F_1 and the corresponding top–down weights;

(AD2) The transmitted signal decays such that the longer the delay the larger the signal loss.

The second assumption requires that delay adaptation occurs, at least in part, at the dendrites. Here we restrict our attention to passive dendrites. For example, the cable equation for a linear RC-membrane cylinder suggests that the amplitude of the voltage response at the soma falls exponentially as a function of the distance between the soma and dendritic synapse ([35]).

In view of these considerations, the input to F_2 from F_1 becomes

$$T_j(t) = D \sum_{1\le i\le m} z_{ij}(t) f_1(x_i(t-\tau_{ij}(t)))g_\alpha(\tau_{ij}), \tag{4.1}$$

where α is a parameter measuring the signal decay rate, $g_\alpha(\tau)$ is a continuous monotone decreasing function of τ with the properties that $g_\alpha(0) > 0$ and $g_\alpha(\tau)$ with a fixed positive τ can be made arbitrarily small if α is sufficiently large. A simple example, that will be used in what follows, is $g_\alpha(\tau) = e^{-\alpha\tau}$ though the decay rate derived from the cable equation for a linear RC membrane cylinder is slightly more complicated.

The equation that describes the delay adaptation is

$$\beta\frac{d\tau_{ij}(t)}{dt} = -\left(\tau_{ij}(t) - \tau_{ij}^0\right) + E\left[1 - h_{ij}(t)\right], \tag{4.2}$$

where τ_{ij}^0 is the fixed anxonal delay between the corresponding F_1 and F_2 neurons, $h_{ij}(t)$ represents the output of the hidden layer of neurons and describes the similarity between the output signal from F_1 and the corresponding component of the top–down weight.

In addition, the LTM equation for the adaptation of the top–down weights must be also modified so that the template will point to the direction of the delayed and exponentially decayed output from the F_1 layer, i.e.

$$\gamma\frac{dz_{ji}(t)}{dt} = f_2(x_j(t))\left[-z_{ji}(t) + f_1(x_i(t-\tau_{ij}))e^{-\alpha(\tau_{ij}(t)-\tau_{ij}^0)}\right]. \tag{4.3}$$

Equations (4.1)-(4.3), (3.1)-(3.2) and (3.4), together with the PART reset mechanism describe the dynamics of the PART-D network in terms of a system of functional differential equations in which the delay is adaptive.

Much of PART-D network's computational performance, described by the long-term and transient behaviors of the model equations, is similar to those of PART, except a minor change of the transient learning formula, that plays an essential role in the PART-D algorithm for data clustering:

$$\lim_{\delta\to 0} z_{iJ}^{\epsilon_1,\epsilon_2,\delta}(1) = \begin{cases} 0 & \text{if } h_{iJ}(0)=0, \\ \frac{L}{L+l_i} & \text{if } h_{iJ}(0)=1, \end{cases}$$

and

$$\lim_{\epsilon_1\to 0,\beta\to 0} z_{Ji}^{\epsilon_1,\epsilon_2,\delta}(1) = (1-q)z_{Ji}(0) + qf_p(I_i)e^{-\alpha(\tau_{iJ}^*+\tau_{iJ}^0)},$$

where $q = 1 - e^{-1/\gamma}$.

This fast learning formula gives the basis for the learning rule (to update all synaptic weights to and from the candidate cluster neuron) in the PARD-D algorithm.

Note that PART is a special case of PART-D if we choose $\alpha = \infty$. This describes the essential difference between PART and PART-D in the decision whether two points belong to the same projective cluster: In PART, information in the subspace where two points are apart from each other play no role, while in PART-D this information plays a complementary role. This is particularly important in the case where the information obtained from the projective subspace is insufficient to decide a winning cluster neuron. To illustrate this, let us consider the case where before an input is presented to the network, at least two F_2 cluster neurons A and B are committed and the corresponding projective subspaces have $|X|$ and $|Y|$ dimensions. Thus, the initial values of the bottom-up weights from these subspaces to the committed neurons are $L/(L-1+|X|)$ and $L/(L-1+|Y|)$, respectively. Assume that within subspaces of dimensions s and d of these projective subspaces, the new input is close to the points in the two projective clusters. Then the total net inputs to cluster neurons A and B are given, in PART, by

$$T_A = \frac{DLs}{L-1+|X|}, \qquad T_B = \frac{DLd}{L-1+|Y|}.$$

If these two net inputs are distinguishable, say $T_A = T_B$(this can happen, for example, $L=2, |X|=3, s=2, |Y|=7, d=4$), then PART cannot select a winning neuron, and will randomly pick up one. On the other hand, PART-D calculates the net inputs as, assuming axonal delays are the same constant τ,

$$T_A = \frac{DLse^{-\alpha\tau}}{L-1+|X|} + \frac{DL(|X|-s)e^{-\alpha(E+\tau)}}{L-1+|X|},$$

$$T_B = \frac{DLde^{-\alpha\tau}}{L-1+|X|} + \frac{DL(|Y|-d)e^{-\alpha(E+\tau)}}{L-1+|Y|},$$

here the dimensions $|X|-s$ and $|Y|-d$, with respect to which the new input and the features of the two projective clusters are dissimilar, do play a critical role since the first term of T_A and T_B is the same. It turns out that this complementary role ensures that if $d > s$ then node B is a winner. This is of course natural, since the input pattern shares the same feature of the cluster represented by neurons A and B, in $s-$ and $d-$ dimensional subspaces.

5 Other developments and applications

5.1 PART for categorical data and applications to commercial data. A modification of PART architecture was proposed in [28] to cluster high dimensional categorical data arising from many practical problems such as marketing analysis. The modification of the architecture is provided in order to deal with categorical attributes, a sample example of such a data set is given in Table 6 below.

Tablo 6 A sample categorical data set.

Records	Values
$\mathbf{x}_1$	(A,A,A,A,B,B)
$\mathbf{x}_2$	(A,A,A,A,C,D)
$\mathbf{x}_3$	(A,A,A,A,D,C)
$\mathbf{x}_4$	(B,B,C,C,D,C)
$\mathbf{x}_5$	(B,B,D,D,C,D)

The modified algorithm, carrying the functional name PARTCAT (PART for CATegorical data), follows the similar architecture to PART and with the principal difference between PARTCAT and PART in how the top-down weights are updated in the learning phase.

To be precise, we need to introduce the so called *symbol table and frequency table* for a given categorical data, which gives certain numerical characterizations of categorical records. Given a d-dimensional categorical data set D, let A_j be the categorical variable of the j-th dimension$(1 \le j \le d)$. We define its domain by $DOM(A_j) = \{A_{j1}, A_{j2}, ..., A_{jn_j}\}$ and we call $A_{jr}(1 \le r \le n_j)$ a state of the categorical variable A_j. Then a symbol table T_s of the given data set is defined as:

$$T_s = (\mathbf{s}_1, \mathbf{s}_2, ..., \mathbf{s}_d),$$

where $\mathbf{s}_j$ is a vector defined as $\mathbf{s}_j = (A_{j1}, A_{j2}, ..., A_{jn_j})^T$. Note that the symbol table for a data set is not unique, and one such table for the data set in Table 6 is given below:

Table 7 One of the symbol tables of the data set in Table 6.

$$\begin{pmatrix} A & A & A & A & B & B \\ B & B & C & C & C & C \\ & & D & D & D & D \end{pmatrix}$$

The frequency table of a cluster is computed according to a symbol table of that data set and it has exactly the same dimension as the symbol table. Let C be a cluster, then the frequency table $T_f(C)$ of cluster C is defined as

$$T_f(C) = (\mathbf{f}_1(C), \mathbf{f}_2(C), ..., \mathbf{f}_d(C)), \tag{5.1}$$

where $\mathbf{f}_j(C)$ is a vector defined as

$$\mathbf{f}_j(C) = (f_{j1}(C), f_{j2}(C), ..., f_{jn_j}(C))^T, \tag{5.2}$$

where $f_{jr}(C)(1 \le j \le d, 1 \le r \le n_j)$ is the number of data points in cluster C which take value A_{jr} at the jth dimension.

For a given symbol table of the data set, the frequency table of each cluster is uniquely determined. For example, for the data set in Table 6, let $C = \{\mathbf{x}_1, \mathbf{x}_2, \mathbf{x}_3\}$

Table 8 The frequency table computed from the symbol table given in Table 7.

$$\begin{pmatrix} 3 & 3 & 3 & 3 & 1 & 1 \\ 0 & 0 & 0 & 0 & 1 & 1 \\ & & 0 & 0 & 1 & 1 \end{pmatrix}$$

be a cluster, if we use the symbol table presented in Table 7, then the corresponding frequency table for the cluster C is given below:

In PARTCAT, the selective output signal from F_1 node v_i to F_2 node v_j is defined as

$$h_{ij} = h(x_i, z_{ij}, z_{ji}) = \delta(x_i, z_{ji}) l_0(z_{ij}), \tag{5.3}$$

where $\delta(\cdot, \cdot)$ is the Simply Matching distance [5], i.e.

$$\delta(a, b) = \begin{cases} 1, & \text{if } a = b, \\ 0, & \text{if } a \neq b. \end{cases} \tag{5.4}$$

The learning formula of PARTCAT is essentially the same as that of PART. However, the top-down weights must be updated in a different fashion due to the categorical nature of the attributes. In particular, for the learning rule of top-down weight z_{ji}^{new} of the committed winning F_2 node v_j which has passed the vigilance test, we need to change it so that it is suitable for categorical values. To do this, let T_s be a symbol table of the input data set and $T_f(C_j)$ be the frequency table for F_2 node v_j, where C_j is the cluster in node v_j. Let $f_{kr}(C_j)$ be the number of elements in cluster C_j whose kth attribute takes value A_{kr}, i.e.

$$f_{kr}(C_j) = |\{\mathbf{x} \in C_j : \mathbf{x}_k = A_{kr}\}|,$$

where A_{kr} is a state of the kth variable's domain $DOM(A_k) = (A_{k1}, A_{k2}, ..., A_{kn_k})$.

Let N_j denote the number of elements in F_2 node v_j, then we have

$$N_j = \sum_{r=1}^{n_k} f_{kr}(C_j), \quad \forall k, j.$$

Now we can define the learning rule for top-down weight z_{ji}^{new} of the committed F_2 node v_j to v_i as

$$z_{ji}^{new} = A_{il_i}, i = 1, 2, ..., m,$$

where $l_i (1 \leq l_i \leq n_i)$ is defined by

$$l_i = \arg \max_{1 \leq r \leq n_i} f_{ir}^{new}(C_j),$$

and $f_{ir}^{new}(C_j)$ is defined by

$$f_{ir}^{new}(C_j) = f_{ir}^{old}(C_j) + \delta(I_i, A_{ir}).$$

Therefore, in PARTCAT, we no longer need the distance parameter σ since we use Simple Matching distance in our algorithm. The PARTCAT algorithm was designed, and its performance was tested on Sun Blade 1000 workstation using GUN C++ compiler for three real world data sets (Soybean Data, Zoo Data and Mushroom Data) obtained from UCI Machine Learning Repository [9]. All these data sets have class labels assigned to the instances. In all simulations, we set $L = 2$ and $\theta = 0$ and compare the clustering results via popular k-Modes algorithm[33]. Let us describe one of the simulations (Soybean Data) here for illustrative purpose.

The soybean disease data set [9] has 47 records each of which is described by 35 attributes. Each record is labeled as one of the 4 diseases: diaporthe stem rot, charcoal rot, rhizoctonia root rot and phytophthora rot. Except for the phytophthora rot which has 17 instances, all other diseases have 10 instances each. We use D1, D2, D3, D4 to donote the 4 diseases. Since there are 14 attributes that have only one category, we only selected other 21 attributes for the purpose of clustering. We set the number of initial subspace dimensions ρ_0 to be 7, expected maximum number of clusters to be formed at each clustering level k to be 4, and the minimum members for each cluster N_{min} to be 20. Applying PARTCAT to the soybean data set using these parameters, we got the results described in Table 9, in comparison with Table 10 obtained by applying the k-Modes algorithm.

Table 9 The misclassification/contingency matrix obtained by applying PARTCAT to the soybean data set.

	D1	D2	D3	D4
C1	10	0	1	0
C2	0	10	0	0
C3	0	0	9	0
C4	0	0	0	17

Table 10 The misclassification/contingency matrix obtained by applying the k-Modes algorithm to the soybean data set.

	D1	D2	D3	D4
C1	7	0	0	10
C2	2	10	0	0
C3	1	0	0	7
C4	0	0	10	0

From Table 9 and Table 10, we see that there is only one point which is misclassified by the PARTCAT algorithm, but there are ten points which are misclassified by the k-Modes algorithm.

For each cluster of the soybean data set, the subspace dimensions associated with each cluster found by PARTCAT are described in Table 11. From Table 11, we see that the subspace dimensions associated with different clusters are different.

Table 11 The subspace dimensions associated with each cluster in the soybean data set.

Clusters	Subspace dimensions
C1	2, 3, 11, 16, 18, 19, 21
C2	2, 3, 8, 11, 13, 14, 15, 16, 17, 18, 19, 20, 21
C3	3, 7, 15, 18, 19, 20
C4	2, 7, 11,14, 15, 17, 18, 19, 20, 21

We did apply PARTCAT algorithm to a data set obtained from a large bank in Canada for the performance analysis of all branches of the bank, the clustering results seem to be quite satisfactory.

5.2 Application 1: Gene expression. In [47], PART was used as a gene filtering method for the construction of robust prognostic predictors. Subspace clustering is essential for establishing prognostic predictors of various diseases using DNA microarray analysis technology, since it is desired to selectively find significant genes for constructing the prognostic model and also necessary to eliminate nonspecific genes or genes with error before constructing the model. According to the authors,

> "Genes selected by PART were subjected to our FNN-SWEEP modeling method for construction of a cancer class prediction model. The model performance was evaluated through comparison with a conventional screening signal-to-noise (S2N) method or nearest shrunken centroids (NSC) method. The FNN-SWEEP predictor with PART screening could discriminate classes of acute leukemia in blinded data with 97.1% accuracy and classes of lung cancer with 90.0% accuracy, while the predictor with S2N was only 85.3% and 70.0% or the predictor with NSC was 88.2% and 90.0%, respectively. The results have proven that PART was superior for gene screening."

5.3 Application 2: Neural spiking trains clustering. In [36], PART was used as an effective tool for clustering neural spiking trains via transient behaviors. It was noted that "the detection of nonstationarities in neural spike trains recorded from chronically implanted multielectrode grids, such as transient synchronizations in a neural sub- population, becomes increasingly difficult as the number of electrodes increases". This calls for unsupervised learning algorithms that can be used to "group, or cluster, spike trains based on the presence of local, shared features". The feature of PART that allows comparisons be made between inputs and learned patterns using a subset of the total number of spikes available enables the network to learn the characteristics that defines each cluster making as few assumptions about the statistical properties of the spike trains as possible.

> "The result is an extremely powerful tool for clustering neural spike trains that is computationally inexpensive. The fact that projective clustering dramatically increases the ability of an artificial neural network to discover patterns in its sensory inputs raises the question of whether analogous mechanisms operate in the nervous system. Thus we anticipate that PART neural networks will not only have increasing applications for data analysis, but also have the potential to provide insights into the computational activities of the nervous system."

5.4 IMC-PART: Order independence clustering. In PART, like ART and many other neural networks based on clustering algorithms, the clustering results depend on the order in which the data set is presented. To overcome this shortcome, in [14] and [11], we developed a simple clustering algorithm - Iterative Mean Clustering (IMC) and the related dynamical system theory to show why this algorithm produces meaningful clustering results independent of the data input order. This algorithm, similar in spirit to the statistical method *mean-shifting*, seems to work extremely well for the case where the data set has one-cluster and the data distribution is unimodal: the algorithm converges to the center of the unique cluster starting from an arbitrary initial value. In a multi-clusters case where the data distribution is multimodal, this algorithm converges to the center of a cluster that is close to the initial value. Therefore, the centers of all clusters of a data set can be found by running this algorithm with several distinct initial values.

This algorithm can be implemented through a ART-like incremental learning net, called *IMC-ART*, and this implementation permits the coexistence of multiple winners. Such an IMC algorithm can also be incorporated into the PART architecture to generate the so-called *PART-A algorithm*, which has the desirable order-independence property for subspace clustering.

The key to such a success in IMC-PART lies in the close relation between the clustering process and the iteration of a discrete map. Vaguely, here are the essential steps of IMC-PART:

1. Select a parameter σ;
2. Choose an initial data point x_0;
3. For $i = 1, 2, \cdots$, set x_i to be the mean of all data points whose distances to x_{i-1} are less than or equal to σ.

To investigate the convergence of such an iteration, for a given data set $S \subset \mathbb{R}^m$ and for a given $x \in \mathbb{R}^m$, let

$$A_{S,\sigma}(x) = \{y \in S : ||y - x|| \leq \sigma\},$$

and define $T_{D,\sigma} : \mathcal{D}(T_{S,\sigma}) \subset \mathbb{R}^m \to \mathbb{R}^m$ by

$$T_{S,\sigma}(x) = \frac{1}{|A_{S,\sigma}(x)|} \sum_{y \in A_{S,\sigma}(x)} y,$$

where

$$\mathcal{D}(T_{S,\sigma}) = \{x \in \mathbb{R}^m : A_{S,\sigma}(x) \neq \varnothing\},$$

and $|A_{S,\sigma}(x)|$ denotes the number of elements in the set $A_{S,\sigma}(x)$.

Let $f : \mathbb{R}^m \to \mathbb{R}$ denote the ddf of S. Define $T : \mathcal{D}(T) \subset \mathbb{R}^m \to \mathbb{R}^m$ by

$$T(x) = \frac{\int_{||y-x|| \leq \sigma} y f(y) dy}{\int_{||y-x|| \leq \sigma} f(y) dy},$$

where

$$\mathcal{D}(T) = \{x \in \mathbb{R}^m : \int_{||y-x|| \leq \sigma} f(y) dy \neq 0\}.$$

A ddf f is said to be *unimodal* if it has only one local maximum, *bimodal* if it has two local maxima, *multimodal* if it has more than one local maxima. In [11], it was proved that for a unimodal ddf f, T has a unique fixed point in $\mathcal{D}(T)$, and the sequence $\{x_n\}_{n=0}^{\infty}$ defined by $x_n = T(x_{n-1})$ converges to the fixed point for an arbitrary initial value $x_0 \in \mathcal{D}(T)$.

5.5 Clusters splitting and bifurcation. The multimodel case is more complicated, and the splitting of a cluster is intimately related to the bifurcation theory. Let f be a multimodal density function with finite many local maxima and minima, and *supp*$(f) = R$. Let $r > 0$ be the minimum distance between a local maximum and an adjacent local minimum of f. Let $0 < \sigma < r/2$. Then

- For every given local maximum p of f, T has a unique fixed point in $(p - \sigma, p + \sigma)$ and this fixed point is asymptotically stable;
- For every given local minimum v of f, T has a unique fixed point in $(v - \sigma, v + \sigma)$ and this fixed point is unstable;

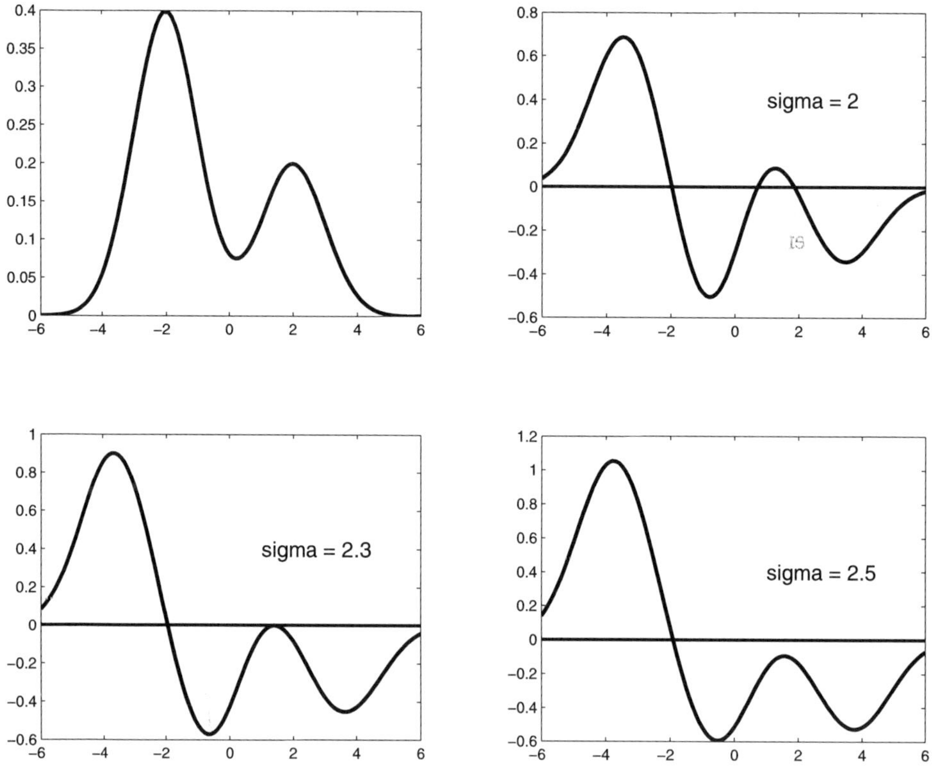

Figure 5 An example of cluster splitting: a data set whose ddf is a mixture of two Gaussian distributions with different sizes (top-left). The size of a cluster center σ turns out to be a natural bifurcation parameter for a related iterative map, $\sigma = 2.3$ is a bifurcation value where the map changes from having two stable fixed points to having one stable fixed points.

- T has no other fixed points.

It is very interesting to see how the performance is affected by the parameter σ and noise when the data consists of at least two clusters, to discuss this in a little detail, let us introduce

Definition 5.1 Let x^* be a stable fixed point of T. We call the domain of attraction of x^* given by

$$C(x^*) = \{x_0 \in S : \lim_{n \to \infty} T^n(x_0) = x^*\}$$

the cluster centered at x^*.

We consider the case the ddf f is bimodal. Depending on the size of σ, we have the following two generic types (bifurcation!):

- Type A (σ is small): T has three fixed points $a < c < b$, where a and b are asymptotically stable and c is unstable. Two clusters centered at a and b can be obtained;
- Type B (σ is large): T has only one zero c, which is asymptotically stable. Only one cluster can be obtained.

Therefore, σ is a bifurcation parameter and it plays a critical role in determining whether a cluster should be further divided. To demonstrate this, [14] provided an example where f is the mixture of two Gaussian distributions with different sizes:

$$f(x) = \frac{1}{\sqrt{2\pi}} e^{-\frac{1}{2}(x+2)^2} + \frac{1}{2\sqrt{2\pi}} e^{-\frac{1}{2}(x-2)^2}.$$

This example shows how the sizes of clusters affect the clustering result. Figure shows that $\sigma = 2.3$ is a critical value: when $\sigma < 2.3$, T has three fixed points, while T has only one fixed point if $\sigma > 2.3$. In this case even when σ is up to 2.0, T has two well-separated asymptotically stable fixed points, therefore two well-separated clusters can be obtained. Note that when $\sigma = 2.5$, the larger cluster has completely absorbed the smaller one. Note that in the figure, the fixed points of T are converted to zeros of a related map.

6 Remarks and open problems

Construction of a nonlinear dynamical system for the purpose of subspace clustering does need some intuitions, but these must be supported by biological, physiological and neuro-scientific evidences and careful development of related mathematical theory. Here, we provide a short discussion of our design of PART architecture for pattern recognition in high dimensional data sets.

Most mathematical results, needed for the development of relevant algorithms, that have been obtained so far are proved only when signal functions are step functions with thresholds. This reflects the on-or-off characteristic of a single neuron, and makes the PART algorithm user friendly in the sense that users are required to pick up the thresholds instead of picking up signal functions. This can be regarded as a reflection of the simple minded "yes" or "no" answer to a very complicated concept of similarity, and as such, it is clearly an approximation of a continuous scale of a similarity measure. It remains to be an interesting problem whether the same dynamical behaviors hold true if we use general signal functions. It is also desirable to extend PART to deal with fuzzy data.

We note that, similar to the traditional ART systems, the "top-down" weights observe a fast-commit/slow-recode dynamics. However, in PART these weights track the inputs rather than tracking a matched pattern, as is characteristic of most ART systems where the matched pattern often tends to code a low dimensional projection of a set of inputs (a major task of the PART algorithm). It would be interesting to investigate how this tracking the input itself affects the convergence of top-down weight. More importantly, examining the alternative dynamic of tracking a matched pattern should be a very useful direction for future investigations.

If variables of input vectors are not independent of each other, we anticipate to find clusters only in skewed subspaces or even submanifolds. Initial progress has been made, see [29], in terms of incorporating the principal component analysis into PART, but much remains to be done.

In summary, viewing subspace clustering of a high dimensional data set as an inverse process of pattern formation in a nonlinear dynamical system to be constructed adaptively from the data set provides a plausible approach towards an important cognitive task of our brains. The foundation has been laid down in ART and the new development of PART and PART-D ties subspace clustering to an essential part (invariant manifold reduction) of nonlinear dynamical system theory, supported by some recent findings and evidence of delay adaptation. This, in turns, calls for theoretical development of delay differential equations with state-dependent delay and new mathematics.

> "In contrast, to do psychology or neuroscience, you needed to discover new intuitions as well as new type of mathematics with which to analyze these intuitions, ..., in order to self-organize intelligent adaptive processes in real time, the brain needs nonlinear feedback processes that describe dynamical interactions among huge numbers of units acting on multiple spatial and temporal scales. Such processes are not easy to think about or to understand. One of the controversies that I experienced early on was whether we needed differential equations at all. Many people wanted discrete or symbolic models, but symbols emerge from their dynamics" (S. Gorssberg, pp194-195, in [6]).

Acknowledgment

I would like to thank Dr. Y. Cao (currently at Boston University) for working with me painfully at the initial stage of developing PART, G. Gan and Professor Z. Yang for the collaboration on PARTCAT with applications to performance analysis of a Canadian Branch data. I would like to thank Professor J. Milton and Dr. J. Hunter for helping me to understand the neurophysiological evidence of the selective signal mechanism from the view point of delay adaptation and neural signal loss due to the time lag in neural signal transmission, and for their collaboration on PART-D and on applications of PART to neural signal spiking trains clustering. I would like to thank the Biotechnology group at Nagoya University led by Dr. H. Honda for discussions of applications of PART for gene expression data analysis. Over the last few years, I have benefited very much from the close collaboration between Laboratory for Industrial and Applied Mathematics at York University and Generation 5 Mathematical Technologies Inc. and I salute the support and appreciation of my basic research by Generation 5 Chief Scientist Dr. W. Huang and President Dr. M. Krneta. My special thanks go to Professor S. A. Campbell for her leadership and unselfish service to a MITACS project which led me to the field of data clustering. Finally, I wish to express my sincere appreciation to Bill (Professor Langford) for his continuing support, thoughtful comments and encouragement to my endeavor in looking at pattern recognition from a nonlinear dynamics point of view.

References

[1] R. Agrawal, J. Gehrke, D. Gunopilos, and P. Raghavan, *Automatic subspace clustering of high dimensional data for data mining applications*, SIGMOD '98, 1998, pp. 94-105.

[2] C. C. Aggarwal, C. Procopiuc, J. L. Wolf, P. S. Yu, and J. S. Park, *Fast algorithms for projective clustering*, SIGMOD '99, 1999, pp. 61-72.

[3] C. C. Aggarwal, and P. S. Yu, *Finding generalized projected clusters in high dimensional spaces*, SIGMOD '00, 2000, pp. 70-81.

[4] H. Agmon-snir, and I. Segev, *Signal delay and input synchronization is passive dendritic structures*, J. Neurophysiol, 70, 1993, pp. 2066–2085.

[5] M. Anderberg, *Cluster Analysis for Applications*, New York: Academic Press, 1973.

[6] J. A. Anderson and E. Rosenfeld, *Talking Nets, An Oral History of Neural Networks*, The MIT Press, Cambridge, 1998.

[7] M. A. Arbib, *The Handbook of Brain Theory and Neural Networks*, The MIT Press, Cambridge, 1995.

[8] K. Beyer, J. Goldstein, R. Ramakrishnan, and U. Shaft, *When Is 'Nearest Neighbors' Meaningful?*, in Proc. 7th International Conference on Database Theory (ICDT '99), 1999, pp. 217-235.

[9] C. Blake and C. Merz, *UCI repository of machine learning databases*, 1998, http://www.ics.uci.edu/ mlearn/MLRepository.html.

[10] Y. Cao, *Neural Networks for Clustering: Theory, Architecture, Algorithm and Applications*, PhD Thesis, York University, 2002.

[11] Y. Cao, *A convergence theorem related to iterative mean clustering*, in Difference and Differential Equations (S. Elaly, J. Ladas, J. Wu and X. Zou eds.), Fields Institute Communications, American Mathematical Society, 2004.

[12] Y. Cao and J. Wu, *Projective ART for clustering data sets in high dimensional spaces*, Neural Networks, vol. 15, 2002, pp. 105-120.

[13] Y. Cao and J. Wu, *Dynamics of projective resonance theory model: the foundation of PART algorithm*, IEEE Trans Neural Networks, vol. 15, 2004, pp. 245-260.

[14] Y. Cao and J. Wu, *IMC-PART neural network for high dimensional data clustering*, preprint, 2005.

[15] G. A. Carpenter and S. Grossberg, *A massively parallel architecture for a self-organizing neural pattern recognition machine*, Computer Vision, Graphics, and Image Processing, vol. 37, 1987, pp. 54-115.

[16] G. A. Carpenter and S. Grossberg, *ART2: Self-organization of stable category recognition codes for analog input patterns*, Applied Optics, vol 26, 1987, pp. 4919-4930.

[17] G. A. Carpenter and S. Grossberg, *ART3: Hierarchical search using chemical transmitters in self-organizing pattern recognition architectures*, Neural Networks, vol. 3, 1990, pp. 129-152.

[18] G. A. Carpenter, S. Grossberg, and J. H. Reynolds, *ARTMAP: Supervised real-time learning and classification of nonstationary data by a self-organizing neural network*, Neural Networks, vol. 4, 1991, pp. 565-588.

[19] G. A. Carpenter, S. Grossberg, and D. B. Rosen, *ART2-A: An adaptive resonance algorithm for rapid category learning and recognition*, Neural Networks, vol. 4, 1991, pp. 493-504.

[20] G. A. Carpenter, S. Grossberg, and D. B. Rosen, *Fuzzy ART: Fast stable learning and categorization of analog patterns by an adaptive resonance system*, Neural Networks, vol. 4, 1991, pp. 759-771.

[21] G. A. Carpenter, S. Grossberg, N. Markuzon, J. H. Reynolds, and D. B. Rosen, *Fuzzy ARTMAP: A neural network architecture for incremental supervised learning of analog multidimensional maps*, IEEE Transactions on Neural Networks, vol. 3, 1992, pp. 698-713.

[22] M. Ester, H.-P. Kriegel, J. Sander, and X. Xu, *A density-based algorithm for discovering clusters in large spatial databases with noise*, in Proc. 2nd International Conference on Knowledge Discovery and Data Mining (KDD '96), Portland, Oregon, 1996, pp. 226-231.

[23] C. W. Eurich, U. Ernst, K. Pawelzik, J. D. Cowan and J.G. Milton, *Dynamics of self-organized delay adaptation*, Phys. Rev. Lett., vol. 82, 1999, pp. 1594-1597.

[24] C. W. Eurich, *Delay adaptation in the nervous system*, Neurocomputing, vol. 38-40, 2000, pp. 741-748.

[25] C. W. Eurich, M. C. Mackey and H. Schwegler, *Recurrent inhibitory dynamics: the role of state dependent distributions of conduction delay times*, J. Theoret. Biol., vol. 216, 2002, pp. 31-50.

[26] B. Everitt, S. Landau, and M. Leese, *Cluster Analysis*, 4th Ed. London: Oxford University Press, 2001.

[27] G. Gan, C. Ma and J. Wu, *A Handbook for Clustering and Pattern Recognition*, preprint, 2004.

[28] G. Gan, Z. Yang and J. Wu, *Subspace clustering for high dimensional categorical data*, preprint, 2004.

[29] A. Gashwer and J. Wu, *Incorporating principal component analysis into PART for projected clustering in skewed subspaces*, preprint, 2004.

[30] S. Grossberg, *Adaptive pattern classification and universal recoding, I: parallel development and coding of neural feature detectors*, Biological Cybernetics, vol. 23, 1976, pp. 121-134.

[31] S. Grossberg, *Adaptive pattern classification and universal recoding, II: feedback, expectation, olfaction, and illusions*, Biological Cybernetics, vol. 23, 1976, pp. 187-202.

[32] R. L. Grossman, K. Chandrika, P. Kegelmeyer, V. Kumer and R. R. Namburu (Eds.), *Data Mining for Scientific and Engineering Applications*. Boston: Kluwer Academic, 2001.

[33] Z. Huang, *Extensions to the k-means algorithm for clustering large data sets with categorical values*, Data Mining and Knowledge Discovery, vol. 2, 1998, pp. 283-304.

[34] L. Hubert and P. Arabie, *Comparing partitions*, Journal of Classification, vol. 2, 1985, pp. 193-218.

[35] J. Hunter, personal communication, 2003.

[36] J. Hunter, J. Wu and J. Milton, Clustering neural spike trains via transient responses, preprint, 2004.

[37] A. K. Jain and R. C. Dubes, *Algorithms for Clustering Data*, Upper Saddle River, NJ: Prentice-Hall, 1988.

[38] A. K. Jain, M. N. Murty, and P. J. Flynn, Data clustering: a review, *ACM Computing Surveys*, 31, 1999, pp. 264-323.

[39] L. Kaufman and P. Rousseeuw, *Finding Groups in Data - An Introduction to Cluster Analysis.* New York: John Wiley & Sons, 1990.

[40] J. Milton, *Dynamics of Small Neural Networks*, Americal Mathematical Society, Providence, R.I., 1996.

[41] J. Milton, J. Hunter and J. Wu, *Self-Supervised delay adaptation for projective clustering in high dimensional spaces*, preprint, 2004.

[42] T. Mundel, J. Milton, A. Dimitrov, H. Wilson, C. Pelizzari, S. Uftring, I. Torres, R. Erickson, J-P. Spire and V. Towle, *Transient inability to distinguish between faces: electrophysiological studies*, J. Clin. Neurophysiol., vol 20, 2003, pp. 102-110.

[43] R. T. Ng and J. Han, *Efficient and effective clustering methods for spatial datu mining*, in Proc. 20th VLDB Conference, Santiago, Chile, 1994, pp. 144-155.

[44] J. Sander, M. Ester, H.-P. Kriegel and X. Xu, *Density-based clustering in spatial databases: the algorithm GDBSCAN and its applications*, Data Mining and Knowledge Discovery, vol. 2, 1998, pp. 169-194.

[45] L. R. Stanford, *Conduction velocity variations minimize conduction time differences among retinal ganglion cell axons*, Science, vol. 238, 1987, pp. 358–360.

[46] B. Stevens, S. Tanner and R. D. Fields, *Control of myelination by specific patterns of neural impulses*, J. Neuroscience, vol. 18, 1998, pp. 9303-9311.

[47] H. Takahashi, T. Kobayashi and H. Honda, *Construction of robust prognostic predictors by using projective adaptive resonance theory as a gene filtering method*, Bioinformatics Advance Access published August 12, 2004.

[48] S. Theodoridis, and K. Koutroumbas, *Pattern Recognition*, Academic Press, New York, 1999.

[49] S. R. Williams and G. J. Stuart, *Dependence of EPSP efficacy on synapse location in neocortical pyramidal neurons*, Sciences, vol. 295, 2002, pp. 1907-1910.

[50] J. R. Williamson, *Gaussian ARTMAP: a neural network for fast incremental learning of noisy multidimensional maps*, Neural Networks, vol. 9, 1996, pp. 881-897.

[51] J. Wu, *Introduction to Neural Dynamics and Signal Transmission Dalay*, Walter de Gruyter, Berlin, 2001.

[52] K. Yeung and W. Ruzzo, *Principal component analysis for clustering gene expression data*, Bioinformatics, vol. 17, 2001, pp. 763-774.

Fields Institute Communications
Volume **49**, 2006

Computation of Limit Cycles – The Second Part of Hilbert's 16th Problem

Pei Yu
Department of Applied Mathematics
The University of Western Ontario
London, Ontario N6A 5B7 Canada
pyu@pyu1.apmaths.uwo.ca

Abstract. In this paper, we study the second part of Hilbert's 16th problem. After an extensive review of the problem, we present two main results obtained recently. The first result is for cubic-order Z_2-equivariant vector fields. Local bifurcation analysis with normal form theory shows that a Z_2-equivariant vector field can have 12 limit cycles, i.e., $H(3) \geq 12$. The second result is for the weakened Hilbert's 16th problem in the study of higher-order perturbed polynomial Hamiltonian systems. Global bifurcation analysis based on the detection function method is used to consider a degree 9, Z_{10}-equivariant perturbed Hamiltonian system. It is shown that such a system can have at least 80 limit cycles, i.e., $H(9) \geq 80$.

1 Introduction

Since David Hilbert [22, 23] presented the well-known 23 mathematical problems to the Second International Congress of Mathematicians in 1900, great advances inspired by the famous address have been achieved in many different mathematical research areas. However, two of the 23 problems remain unsolved after more than one hundred years. One of them is the 16th problem titled as the *Problem of the topology of algebraic curves and surfaces.* This problem includes two parts. In the first part, Hilbert suggested a thorough investigation of the relative positions of the separate branches of algebraic curves in nth-order vector fields, which is in the area of real algebraic geometry. Significant progresses have been made in this area of research [21, 63, 19, 78, 74, 52, 4, 71, 34, 61, 53]. However, this part, a request for an investigation, is a rather open-ended problem and may not be fully resolved [24]. The second part of the problem was outlined by Hilbert as follows [22, 23]:

2000 *Mathematics Subject Classification.* Primary 34C07, 34C23; Secondary 37C27, 37C29.
This work was supported by NSERC Grant #R2686A02.

"*In connection with this purely algebraic problem, I wish to bring forward a question which, it seems to me, may be attacked by the same method of continuous variation of coefficients, and whose answer is of corresponding value for the topology of families of curves defined by differential equations. This is the question as to the maximum number and position of Poincaré's boundary cycles (limit cycles) for a differential equation of the first order and degree of the form*

$$\tfrac{dy}{dx} = \tfrac{Y}{X}, \tag{1.1}$$

where X and Y are rational integral functions of the nth-degree in x and y. Written homogeneously, this is

$$X\left(y\,\tfrac{dz}{dt} - z\,\tfrac{dy}{dt}\right) + Y\left(z\,\tfrac{dx}{dt} - x\,\tfrac{dz}{dt}\right) + Z\left(x\,\tfrac{dy}{dt} - y\,\tfrac{dx}{dt}\right) = 0, \tag{1.2}$$

where X, Y and Z are rational integral homogeneous functions of the nth-degree in x, y and z, and the latter are to be determined as functions of the parameter t."

In this part of the problem, Hilbert asked to consider the upper bound of the number of limit cycles and their relative locations in polynomial vector fields of order n. This part of problem is related to ordinary differential equations and dynamical systems. Generally, this part of the problem is what usually meant when talking about Hilbert's 16th problem.

To be more specific, consider a polynomial vector field in the real (x, y)-plane, described by

$$\dot{x} = P_n(x, y), \qquad \dot{y} = Q_n(x, y), \tag{1.3}$$

where the dot denotes differentiation with respect to time t, and P_n and Q_n are nth-degree polynomials of x and y. More precisely, the statement of second part of Hilbert's 16th problem is given below.

Hilbert's 16th Problem Estimate a uniform upper bound for the number of limit cycles of (1.3), $H(n)$, for any $n \in \mathbb{Z}_+$.

Traditionally, the second part of 16th problem has three subproblems [27].

Problem 1.1 Is it true that a planar polynomial vector field has but a finite number of limit cycles?

Problem 1.2 Is it true that the number of limit cycles of a planar polynomial vector field is bounded by a constant depending on the degree of polynomials only?

Problem 1.3 Give an upper bound for $H(n)$.

These three problems are related in chain. Solving Problem 3 implies both Problems 1 and 2 solved; while solving Problem 2 only means Problem 1 solved. In general, none of the three problems has been completely solved. Problem 1 was solved for each *given individual* polynomial system [30, 16]. Problem 1 is still open even for $n = 2$. In fact, it is only known so far the lower bound of quadratic systems is 4, i.e., $H(2) \geq 4$ [8, 70].

Although it has not been possible to obtain the uniform upper bound for $H(n)$ directly up to now, various efforts have been made in finding the maximal number of limit cycles and raising the lower bound of Hilbert number $H(n)$ for general planar polynomial systems or for individual degree of systems, through which people hope to get close to the estimation of the upper bound of $H(n)$. In this paper, after an extensive review of the problem, we present two main results obtained recently, regarding the lower bound of $H(n)$. First, we will apply the technique of normal form computation [82] to show that a cubic-order Z_2-equivariant vector field can have 12 small limit cycles, i.e., $H(3) \geq 12$ [83, 84, 85, 86, 87, 88]. Second, we will consider the weakened Hilbert's 16th problem for higher-order perturbed polynomial Hamiltonian systems. In particular, global bifurcation analysis with the aid of detection function method is used to investigate a 9th-degree, Z_{10}-equivariant perturbed polynomial Hamiltonian system. It is shown that such a system can have at least 80 limit cycles, i.e., $H(9) \geq 80$ [77].

The rest of the paper is organized as follows. An extensive review is given in the next section. Section 3 is devoted to study cubic-order, Z_2-equivariant vector fields which exhibit 12 limit cycles. In Section 4, the existence of 80 limit cycles is shown in a 9th-degree, Z_{10}-equivariant perturbed Hamiltonian system. Finally, conclusion is drawn in Section 5.

2 Literature review

The centennial history of Hilbert's 16th problem is fairly dramatic. Before Hilbert stated this problem, in the 1880s, Poincaré studied planar polynomial vector fields as soon as he proposed his program for developing the geometric theory of differential equations. He first introduced the notion of limit cycle and proved that *a planar polynomial vector field without saddle connections has only a finite number of limit cycles* [27]. Later, Bendixson [7] extended Poincaré's work and further applied Green's function to establish the relationship between the occurrence of closed trajectory and the divergence of a planar vector field. This connection has been continuously extended and strengthened since then. In the same year as Bendixson's paper was published, Hilbert set forth his famous list of the 23 mathematical problems [22].

2.1 The finiteness problem. It is noticed that for the first three decades of the twentieth century, very few people have studied or reported the 16th problem, except for the French mathematician Dulac who published his results [12] in 1923. He claimed that he solved Problem 1.1, the finiteness of the problem.

Dulac's Theorem *Let $\dot{x} = f(x)$ be an analytic planar system, then in any bounded region of the plane there is at most a finite number of limit cycles. Also any polynomial planar system has at most a finite number of limit cycles.*

About thirty years later, in 1952, Bautin [6] proved that any quadratic systems have the most three limit cycles. For some time, it was believed that $H(2) = 3$. A few years later, Petrovskii and Landis also claimed the solution to Problem 1.3 [57, 58] that $H(2) = 3$ and $H(3) \leq P_3(n)$ ($P_3(n)$ is a polynomial of degree 3). In the early 60s of the last century, their claim was disproved by Novikov and the authors themselves. Quadratic systems with four limit cycles were constructed in 1979 by Shi [70], and Chen and Wang [8]. Hence, $H(2) \geq 4$.

A big step back for the history of Hilbert's 16th problem happened in the beginning of 1980's, when Ilyashenko [28] found a strikingly simple counter-example to Dulac's proof of 1923. The gap of Dulac's results was stated by Ilyashenko in 1985 [29]. Around the same time, Ilyashenko proved that *polynomial vector fields with nondegenerate singular points on the plane have a finite number of limit cycles.* Chicone and Shafer [9] in 1983 proved the finiteness of bounded graphs for quadratic systems. Bamón [5] in 1985 reported that he proved "*every quadratic polynomial system has a finite number of limit cycles*".

One of the most important results in the history of Hilbert's 16th problem came out in 1991 by Ilyashenko [30] and in 1992 by Écalle [16], who independently proved the finiteness theorem, pioneered by Dulac, *for every individual, given planar polynomial system.* The general idea of Ilyashenko's proof [31] is based on the properties of Poincaré map, the *expandability and extendibility* for polycycles of analytic planar vector fields. The theorem was proved using the argument of contradiction.

Having considered the *finiteness* problem, the next natural step is to ask if an uniformly bounded number $H(n)$ exists. Although this question still remains open even for the quadratic polynomial vector field, the problem has influenced significantly on different areas of mathematical research, such as geometric theory of polynomial vector fields, bifurcation theory, normal form and algebraic geometry. Considerable developments have been reported in these areas and their intersections [11, 47, 68, 67, 81, 65, 35].

Due to the difficulties in finding this upper bound number for general systems, several simplified versions of Hilbert's 16th problem have been introduced to approach the problem.

2.2 The Abel equation. The Abel differential equation [50, 92] is given by

$$\tfrac{dy}{dx} = f_0(x) + f_1(x)y + f_2(x)y^2 + \cdots + f_n(x)y^n, \tag{2.1}$$

where the coefficients f_j $(j = 0, 1, 2, \ldots, n)$ are continuous functions of x. The general problem is to get an upper bound on the number of limit cycles of equation (2.1). For small values of n, it is expected that the upper bound depends upon n only. In 1980, Shahshahani stated that [69]: For $n \geq 3$, the number of limit cycles of (2.1) is no greater than n. However, the explosion starts at $n = 4$. In fact, Lins [45] found that for $n \geq 4$, the Abel equation might have an arbitrary number of limit cycles. Moreover, there are no specific restrictions on the Poincaré map of Abel equations [55]. In order to obtain the upper bound on the number of limit cycles, the simplified version of Hilbert's 16th problem for Abel equation is stated as follows.

Problem 2.1 Find an upper bound on the number of limit cycles of the Abel equation (2.1) with which the coefficients f_j are trigonometric polynomials of degree no greater than m. The bound should be expressed through n and m only.

This problem is unsolved even for $m = 1$ [27].

2.3 The Liénard system. The Liénard system [44] is defined by the following equations:

$$\dot{x} = y - F_n(x), \quad \dot{y} = -x, \tag{2.2}$$

where $n = 2k + 1$ is odd and the polynomial $F_n(x) = x^n + \sum_1^{n-1} a_j x^j$. The transformation along the y-axis may set the free term of F_n as zero. Therefore, the origin (0, 0) is the only singular point, also the only one source of limit cycles. The studies by Lins *et al.* [46] showed that all the orbits at infinity are repelled by the so called "repelling node", implying that there are no limit cycles at infinity for this system. They also proved that when F_n is cubic, system (2.2) has exactly one limit cycle. Later, Rychkov showed that the Liénard system (2.2) has at most two limit cycles with $F_n(x) = a_1 x + a_3 x^3 + a_5 s^5$. A more general result was reported by Perko [56] in the following theorem.

Theorem 2.2 *For $\epsilon \neq 0$ sufficiently small, system (2.2) with $F_n(x) = \epsilon\,(a_1 x + a_2 x^2 + \cdots + a_{2m+1}x^{2m+1})$ has at most m limit cycles; furthermore, for $\epsilon \neq 0$ sufficiently small, this system has exactly m limit cycles which are asymptotic to circles of radius r_j, $j = 1, 2, \ldots, m$, centered at the origin as $\epsilon \to 0$ if and only if the mth-degree equation*

$$\tfrac{a_1}{2} + \tfrac{3a_3}{8}\,\rho + \tfrac{5a_5}{16}\,\rho^2 + \tfrac{35a_7}{128}\,\rho^3 + \cdots + \binom{2m+2}{m+1}\tfrac{a_{2m+1}}{2^{2m+2}}\,\rho^m = 0$$

has m positive roots $\rho = r_j^2$, $j = 1, 2, \ldots, m$.

This theorem allows us to construct polynomial systems with desired number of limit cycles. As an example, 2 and 3 limit cycles can be obtained from system (2.2) with $F_n(x) = \epsilon\,(72x - 292x^3/3 + 224x^5/5 - 128x^7/35)$ at different perturbation values of ϵ. Here, $m = 3$. The case of 3-limit cycles shows that the inside one at $r = 1$ and the outside one at $r = 3$ are stable, while the one in the middle at $r = 2$ is unstable.

Smale [72] in his "*Mathematical Problems for the Next Century*" described a simplified version of Hilbert's 16th problem regarding the Liénard equation.

Problem 2.3 [72] Find an upper bound on the number of limit cycles for the Liénard system (2.2) through the degree n.

Smale stated that "*The limit cycles of (2.2) are precisely the fixed point of P (Poincaré map). In various talks I raised the question of estimating the number of these fixed points (via some new kind of fixed point theorem)?*"

It should be clarified that until now, no upper bound has even been found for the Liénard equation (2.2).

2.4 $H(2)$: The Dumortier-Roussarie-Rousseau program. After the finiteness theorem in quadratic vector fields was proved by Bamón [5], Roussarie [64] proposed a program for obtaining the existence of $H(2)$. His work is under the spirit of Bautin's work [6], where Bautin introduced the notion of finite cyclicity for bifurcation and proved that a center or fine focus of a quadratic system has cyclicity at most three. Dumortier, Roussarie and Rousseau reduced the problem to study limit cycles bifurcating from polycycles. Their program also proved that limit periodic sets have finite cyclicity. Later they tried to realize Roussarie-Roussarie-Rousseau's program [13, 14] and presented an impressive list of 121 graphics from all possible limit periodic sets of four classes [66] that occur in quadratic systems and proved that if each of the graphics on this list has finite cyclicity, then there exists a uniform upper bound for the number of limit cycles over the class of vector fields. Later the open cases are reduced down to four elementary graphics, but more others [15]. However, as Yakovenko commented on this work, "*all this is half of the story*". New methods are expected to prove the existence of $H(2)$.

2.5 The Hilbert-Arnold problem. In the middle of 1980s, Arnold [1] proposed to consider generic families of smooth vector fields as the first step toward understanding families of analytic vector fields. A smooth analog of the global finiteness conjecture is as follows [3]:

Problem 2.4 *Prove that in a generic finite-parameter family of smooth vector fields on the two-sphere $\mathbb{S}^2$ with a compact parameter base B, the number of limit cycles is uniformly bounded over all parameter values.*

Definition 2.5 A polycycle γ of a vector field is a cyclically ordered collections of equilibrium points $p_1, \ldots, p_k$ (with possible repetitions) and arcs $\gamma_1, \ldots, \gamma_k$ (distinct integral curves consisting possibly of equilibrium points) connecting them in a specific order: the jth arc connects p_j with p_{j+1}, for $j = 1, \ldots, k$.

Definition 2.6 A singular point of a planar vector field is called elementary if the linearization of the field at this point has at least one nonzero eigenvalue. A polycycle is called elementary if it contains elementary singular points only.

Definition 2.7 Denote by $E(k)$ the maximal number of limit cycles that may bifurcate from an elementary polycycle in a typical k-parameter family of smooth planar vector fields (Warning: here k is the number of parameters).

In 2001, Kaloshin proved the following estimation of $E(k)$ [32].

Theorem 2.8 *For any* k, $E(k) \leq 2^{25k^2}$.

This estimation is one of the first upper bound of Hilbert type numbers up to now. But obviously, this form is relatively too high.

2.6 The weakened Hilbert's 16th problem. Consider a polynomial perturbation onto a Hamiltonian vector field:

$$\begin{aligned} \dot{x} &= \frac{\partial H}{\partial y} - \epsilon\, P(x,y), \\ \dot{y} &= -\frac{\partial H}{\partial x} - \epsilon\, Q(x,y), \end{aligned} \tag{2.3}$$

where H is a polynomial Hamiltonian function, P and Q are polynomials, and $0 < \epsilon \ll 1$. We assume that the Hamiltonian function at level $h \in (h_1, h_2)$, i.e., $H(x,y) = h$ contains at least one family of closed orbits γ^h. When $\epsilon = 0$, the integrable Hamiltonian vector field (2.3) does not have any limit cycles. All the cycles are nonisolated, either extend to infinity or end at centers or equilibrium curves. However, when perturbation is added, energy is accumulated inside families of closed orbits. If there is no energy dissipation, i.e., the Abelian integral around the closed orbit γ^h,

$$I(h) = \oint_{\gamma^h} P(x,y)\,dx + Q(x,y)\,dy = 0 \quad (\gamma \subseteq H(x,y) = h), \tag{2.4}$$

limit cycles will be generated from the family of closed orbits γ^h. The integral $I(h)$ is called Poincaré-Melnikov Abelian integral, which is a multi-valued function corresponding to different ovals at the same Hamiltonian level h with fixed polynomials H, P and Q. The close relation between this Abelian integral and the number of limit cycles can be described in the following criterion [62].

Pontryagin Criterion If an oval $\gamma(h)$ of the polynomial H generates a limit cycle of (2.3), then $I(h) = 0$. On the other hand, if $I(h) = 0$ and $I'(h) \neq 0$, then the oval $\gamma(h)$ generates a limit cycle of (2.3).

This criterion combined with the results obtained by Ye [79] and Zhang [90] gives rise to the following Poincaré-Pontryagin-Andronov theorem.

Poincaré-Pontryagin-Andronov Theorem *The following statements hold.*

- *If* $I(h^*) = 0$ *and* $I'(h^*) \neq 0$, *then there exists a hyperbolic limit cycle* L_{h^*} *of system (2.3) such that* $L_{h^*} \to \Gamma_{h^*}$ *as* $\epsilon \to 0$; *and vice versa, if there exists a hyperbolic limit cycle* L_{h^*} *of system (2.3) such that* $L_{h^*} \to \Gamma_{h^*}$ *as* $\epsilon \to 0$, *then* $I(h^*) = 0$ *and* $I'(h^*) \neq 0$.

- *If $I(h^*) = I'(h^*) = I''(h^*) = \cdots = I^{k-1}(h^*) = 0$, and $I^k(h^*) \neq 0$, then system (2.3) has at most k limit cycles for sufficient small ϵ in the vicinity of Γ_{h^*}.*
- *The total number of isolated zeros of Abelian integral (2.4) (taking into account of their multiplicity) is an upper bound number of limit cycles of system (2.3) that bifurcate from the periodic orbits of a period annulus of Hamiltonian system* $(2.3)_{\epsilon=0}$.

The weakened Hilbert's 16th problem introduced by Arnold [2] can now be formulated as follows, which is also called tangential or infinitesimal Hilbert's 16th problem.

The Weakened Hilbert's 16th Problem Find an upper bound of the number of real zeros of the integral (2.4). The bound should be uniform with respect to any collection of polynomials $P \in R[x,y]$ and $Q \in R[x,y]$ of degree $\le n$, H of degree $\le n+1$, and the families of closed orbits γ. It should depend on the degree n only.

According to Poincaré-Pontryagin-Andronov theorem, the weakened problem is concerning not only the number of zeros of the Abelian integral from the perturbed polynomial Hamiltonian system, but also an upper bound of the number of limit cycles bifurcated from the system.

More generally, according to Melnikov theory [56], the number of limit cycles for a perturbed Hamiltonian system is closely related to the number of zeros of the displacement function

$$d(h,\epsilon) = \epsilon\, M_1(h) + \epsilon^2 M_2(h) + \cdots + \epsilon^k M_k(h) + \ldots, \tag{2.5}$$

where the displacement function $d(h,\epsilon)$, defined by the displacement of one section to the flow in Poincaré map, is the function of the Hamiltonian level h and the perturbation number ϵ. The function $M_k(h)$ is called kth-order Melnikov integral. The Abelian integral $I(h)$ mentioned before is equivalent to the first order Melnikov integral $M_1(h)$. It is clear that searching the number of zeros of Melnikov integrals gives the solution of maximal number of limit cycles, and higher-order Melnikov integrals have more isolated zeros and, respectively produces more limit cycles than the first order, the Abelian integral. However, so far, the Melnikov integral is only used up to quadratic order and to cubic order in the symmetric cases. In general the order of Melnikov integral which generates a module of Abelian integrals of a maximal possible dimension is unknown, and its determination appears to be a difficult task involving the solution of the corresponding center-focus problem and viewing the structure of the related center manifold. On the other hand, calculation of the higher-order Melnikov integrals depends on the relative cohomology decomposition of polynomial one-forms and other functions, which is also pending to be further developed [25, 17].

In 1984, the general existential finiteness theorem for the weakened problem was proved by Khovanskii [33], Varchenko [73] and later improved by Novikov and Yakovenko [51]. We give their main results in the following two theorems.

Theorem 2.9 *For any $n \in \mathbb{Z}_+$, the number of isolated zeros of Abelian integrals is uniformly bounded over all Hamiltonian and forms of degree $\le n$.*

Theorem 2.10 *For a generic polynomial $H = H(x,y)$ and an arbitrary differential 1-form $\omega = P(x,y)\,dx + Q(x,y)\,dy$ with polynomial coefficients of degree*

$\leq n$, the number of ovals of the foliation $H = const$, which yield the zero value of the complete Abelian integral $I(t) = \oint_{H=t} \omega$, grows at most as $e^{O_H(n)}$ as $n \to \infty$, where $O_H(n)$ depends only on H.

Now the question is: what is the form of the upper bound number of limit cycles generated from perturbed Hamiltonian systems with respect to the degree n? The answer is still unknown.

2.7 Bifurcation to limit cycle theory. Limit cycle theory takes the central role in Hilbert's 16th problem. Studying the number of limit cycles for differential equations is the most difficult part of the problem. The phenomenon of limit cycle was first discovered and introduced by Poincaré in his four-part article *Integral curves defined by differential equations* [60] published between 1881 and 1886. In this article, Poincaré presented his breakthrough qualitative theory of differential equations which studies the general behaviors of the system without obtaining a specific solution. In order to determine the existence of limit cycles for a given differential equation and the properties of limit cycles, Poincaré introduced the method of topographical system, the successor function, the method of small parameter [59] and the Annular Region Theorem. The famous Poincaré Map is still the most basic tool for studying the stability and bifurcations of periodic orbits.

Bifurcation theory is intimately related to Hilbert's 16th problem. Limit cycles are generated by perturbation via bifurcations. The term originated from Poincaré's article in 1881 was used to describe the "splitting" of equilibrium solutions in a family of differential equations. Physical systems usually have parameters in defining system of differential equations. For example, in the perturbed vector field (2.3), functions $H(x, y)$, $P(x, y)$ and $Q(x, y)$ are polynomials with arbitrary coefficients treated as parameters. In order to understand the qualitative behavior of the system, it is necessary to describe the various bifurcations that occur in the system of differential equations and to determine the parameter values, called bifurcation values. To examine the maximal number of limit cycles for the perturbed vector field (2.3), it is important to consider all different types of bifurcations generating limit cycles in planar vector fields; then to compute the bifurcation values in terms of parameters; and finally to control and determine the parameters which make the system behave as expected. Here the objective is to make the system generating the maximal number of limit cycles.

In general, there are two basic types of bifurcations: local bifurcation and global bifurcation. Limit cycles are generated through bifurcations in many different ways, among them we list several important bifurcations in producing limit cycles:

(i) Multiple Hopf bifurcations from a center or focus;
(ii) Separatrix cycle bifurcations from homoclinic or heteroclinic orbits;
(iii) Global center bifurcations from periodic annuli;
(iv) Limit cycle bifurcations from multiple limit cycles.
(v) Limit cycle bifurcations from graphics with line of singular points.

Limit cycles bifurcated from a focus, center or limit cycle are called local limit cycles or small limit cycles, which are usually studied by normal form theory and other local bifurcation theories (e.g., see [85, 86, 87, 88]). The limit cycles generated from saparatrix cycles, global period annuli or other global structures are called global limit cycles, which are studied by global bifurcation theories, such as Poincaré-Pontryagin-Andronov theorem or higher order Melnikov function analysis [56, 49].

Combinations of local and global bifurcations may allow one to obtain more limit cycles (e.g., see [75]). The complete study of bifurcations of limit cycles should consider all different bifurcations.

2.8 Some lower bound results. While it has not been successful to obtain a uniform upper bound for $H(n)$ directly so far, various efforts have been made in finding the maximal number of limit cycles and raising the lower bound of Hilbert number $H(n)$ for general planar polynomial systems or for individual degree of systems. Lower bound of $H(n)$ for quadratic systems reported by Shi [70], and Chen and Wang [8] is 4, i.e., $H(2) \geq 4$. In 1954, Otrokov [54] obtained the result for any $n \geq 6$, $H(n) \geq (n^2 + 5n - 20 - 6(-1)^n)/2$ by considering a small neighborhood of a singular point. Ilyashenko [26] proved $H(n) \geq (n-1)(n-2)/2$ in 1969. A better estimation came from Christopher and Lloyd in 1995 [10] that $H(2^k - 1) \geq 4^{k-1}(2k - (35/6)) + 3.2^k - (5/3)$ for $n = 2^k - 1$, which was recently corrected by Li [35] as $H(2^k - 1) \geq 4^{k-1}(k - (13/6)) + 2^k - (1/3)$. Several results for lower bounds of specific degrees of systems were also reported. For cubic systems, the best result obtained by Yu *et al.* [83, 86, 87, 88] using normal form theory is $H(3) \geq 12$. For fourth-order systems, Zhang *et al.* [89] recently reported $H(4) \geq 15$ by adding a 4th-degree polynomial perturbation onto a cubic system via bifurcation theory and qualitative analysis. Perturbed equivariant Hamiltonian systems of degrees 5 and 7 have been studied by J. Li *et al.* [36, 42] with the application of the detection function method via numerical computation, yielding $H(5) \geq 24$ and $H(7) \geq 49$.

3 Local problem – 12 limit cycles in cubic-order system

When Hilbert's 16th problem is restricted to the neighborhood of an isolated fixed point, the question is reduced to studying degenerate Hopf bifurcations. For cubic-order systems, 11 limit cycles have been found from two cubic systems [35, 91]. Recently, the following cubic order system:

$$\begin{aligned} \dot{x} &= a\,x + b\,y + a_{30}\,x^3 + a_{21}\,x^2\,y + a_{12}\,x\,y^2 + a_{03}\,y^3, \\ \dot{y} &= b\,x + a\,y + b_{30}\,x^3 + b_{21}\,x^2\,y + b_{12}\,x\,y^2 + b_{03}\,y^3, \end{aligned} \tag{3.1}$$

has been investigated to study the existence of various numbers of small limit cycles [83, 84, 85, 86, 87, 88]. Here, the Jacobian of the system evaluated at the fixed point $(x, y) = (0, 0)$ has eigenvalues $a \pm |b|$. When $|b| > |a|$, the origin is a saddle point; while if $|a| > |b|$, the origin is a node. Further, without loss of generality, assume that

$$a_{03} = -\,b, \quad b_{03} = -\,a \quad \text{and} \quad a_{12} = a, \tag{3.2}$$

under which system (3.1) is symmetric with the origin and has two Hopf-type focus points at $(x, y) = (0, 1)$ and $(0, -1)$. Under the above conditions, 10 small limit cycles have been found from system (3.1) when the origin is a saddle point [20]. Later, it has been further shown that system (3.1) has 12 small limit cycles for a special case [85, 88] as well as for the general Z_2-equivariant case [86, 87].

Note that under the first two conditions of Eq. (3.2), the real part of the eigenvalues of the system evaluated at the two focus points $(0, \pm 1)$ is given by

$$v_0 = \tfrac{1}{2}\,(a_{12} - a), \tag{3.3}$$

if $(a_{12} - a)^2 < 8\,(b^2 - a^2 + b\,b_{12} - a\,a_{12})$. Furthermore, under the condition $a_{12} = a$, the two eigenvalues become a purely imaginary pair if $b\,(b + b_{12}) > 2\,a^2$.

An early developed perturbation method based on multiple scales [82] has been proved computationally efficient in computing the focus values (equivalent to finding normal forms of differential equations). The Maple symbolic program developed in [82] provides a user-friendly tool which can be "automatically" executed in a computer system to find the focus values.

Suppose that the normal form of system (3.1) associated with generalized Hopf singularities, up to $(2k+1)$th order term, is given by

$$\dot{\rho} = \rho\,(v_0 + v_1\,\rho^2 + v_2\,\rho^4 + \cdots + v_k\,\rho^{2k}), \tag{3.4}$$

$$\dot{\theta} = 1 + \dot{\phi} = 1 + t_1\,\rho^2 + t_2\,\rho^4 + \cdots + t_k\,\rho^{2k}, \tag{3.5}$$

where v_0 denotes the term obtained from linear perturbation (see Eq. (3.3)). The basic idea of finding k small limit cycles around the origin is as follows: First, find the conditions such that $v_0 = v_1 = v_2 = \cdots = v_{k-1} = 0$, but $v_k \neq 0$, and then perform appropriate small perturbations to prove the existence of k limit cycles.

3.1 Sufficient conditions for the existence of small limit cycles. In the following, we give two theorems which can be used to prove the existence of small limit cycles. Without loss of generality we may assume $v_k > 0$. Then, solving ρ^2 from the equation $\frac{d\rho}{dt} = 0$ is equivalent to finding the roots of the equation:

$$y^k + c_{k-1}\,y^{k-1} + \cdots + c_1\,y + c_0 = 0, \tag{3.6}$$

where $y = \rho^2$ and $c_{i-1} = \frac{v_{i-1}}{v_k}$, for $i = 1, 2, \ldots, k$. Since we are interested in small amplitude solutions, we may assume that the roots can be put in the form of $y = \rho^2 = O(\epsilon)$ $(0 < \epsilon \ll 1)$. The following two theorems provide sufficient conditions for the existence of small limit cycles. The proofs can be found in [86, 87].

Theorem 3.1 *Let $\beta_0, \beta_1, \ldots, \beta_{k-1}$ be constants such that the following equation:*

$$w^k + \beta_{k-1}\,w^{k-1} + \cdots + \beta_1\,w + \beta_0 = 0, \tag{3.7}$$

has k simple positive roots w_i, $i = 1, 2, \ldots, k$. Then for any continuous functions c_i satisfying

$$c_i(\epsilon) = \beta_i\,\epsilon^{k-i} + o(\epsilon^{k-i}), \quad i = 0, 1, \ldots, k-1, \tag{3.8}$$

equation (3.6) has exactly k simple positive roots in the form of $y_i = \epsilon\,w_i + o(\epsilon)$ for sufficiently small $\epsilon > 0$. Therefore, if

$$v_i = \beta_i\,\epsilon^{k-i} + o(\epsilon^{k-i}), \quad i = 0, 1, \ldots, k, \tag{3.9}$$

with $\beta_k \neq 0$, then Eq. (3.6) has exactly k real positive roots (i.e., system (3.4) has exactly k limit cycles) in a neighborhood of the origin for sufficiently small $\epsilon > 0$.

Note that the roots w_j and the coefficients β_j are mutually determined. However, in many cases, v_j depends on k parameters:

$$v_j = c_j(\epsilon_1, \epsilon_2, \ldots, \epsilon_k), \quad j = 0, 1, \ldots, k. \tag{3.10}$$

In this case, the following theorem is more convenient in applications.

Theorem 3.2 *Assume that the condition (3.10) holds and*

$$c_k(0) \neq 0,$$

$$c_j(0) = 0, \quad j = 0, 1, \ldots, k-1,$$

$$\det\left[\frac{\partial(c_0, c_1, \ldots, c_{k-1})}{\partial(\epsilon_1, \epsilon_2, \ldots, \epsilon_k)}(0)\right] \neq 0. \tag{3.11}$$

Then for any given $\epsilon_0 > 0$, there exist $\epsilon_1, \epsilon_2, \ldots, \epsilon_k$ and $\delta > 0$ with $|\epsilon_j| < \epsilon_0$, $j = 1, 2, \ldots, k$ such that Eq. (3.6) has exactly k real positive roots (i.e., system (3.4) has exactly k limit cycles) in a δ-ball with the center at the origin.

3.2 Existence of 12 limit cycles in a special case. In this subsection, we outline the results of 12 limit cycle for a special case, and will discuss the generic case in the next subsection. Consider system (3.1) and assume $|b| > |a|$ under which the origin is a saddle point. In addition to the conditions given in Eq. (3.2), further suppose that the following condition

$$b_{12} = \frac{1+4\,a^2}{2\,b} - b \tag{3.12}$$

holds, then the two symmetric Hopf-type focus points at $(x, y) = (0, \pm 1)$ have the eigenvalues $\lambda_{\pm} = \pm i$. Now, we have *six* free parameters: a, b, a_{21}, b_{21}, a_{30} and b_{30}, which are involved in the expressions of the focus values.

Since the system is symmetric with the two focus points $(0, 1)$ and $(0, -1)$, we only need to consider one of them, say, $(0, 1)$. Thus, introduce the transformation: $x = 2\,b\,u$, $y = 1 + 2\,a\,u - v$ into system (3.1) to obtain

$$\begin{aligned} \dot{u} &= v + 2\,\bar{a}_{21}\,u^2 + 4\,a\,u\,v - \tfrac{3}{2}\,v^2 + 4\,b\,\bar{a}_{30}\,u^3 - 2\,\bar{a}_{21}\,u^2 v - 2\,a\,u\,v^2 + \tfrac{1}{2}\,v^3, \\ \dot{v} &= -u - 4\bar{b}_{21}u^2 + 2(2a^2 - 2b^2 + 1)uv - 8\bar{b}_{30}u^3 + 4\bar{b}_{21}u^2 v - (2a^2 - b^2 + 1)uv^2, \end{aligned} \tag{3.13}$$

where $\bar{a}_{21}$, $\bar{a}_{30}$, $\bar{b}_{21}$ and $\bar{b}_{30}$ are given in terms of a_{21}, a_{30}, b_{21}, b_{30}, a and b. The Jacobian of Eq. (3.13) evaluated at the origin is now in Jordan canonical form.

Apply the Maple program [82] to system (3.13) to obtain the focus values, v_i, explicitly expressed in terms of the system's coefficients: $v_i = v_i(a, b, \bar{a}_{21}, \bar{b}_{21}, \bar{a}_{30}, \bar{b}_{30})$, $i = 1, 2, \ldots, 7$. Since there are six free coefficients, it may be possible to set $v_i = 0$, $i = 1, 2, \ldots, 6$, but $v_7 \neq 0$. Hence, this seems to give some hope that the system might exhibit 14 small limit cycles.

Letting $v_1 = 0$ yields

$$\bar{a}_{30} = \tfrac{1}{3\,b}\left[a\,(2\,\bar{a}_{21} - 1) + \bar{b}_{21}\,(4\,\bar{a}_{21} + 4\,a^2 - 4\,b^2 + 1)\right], \tag{3.14}$$

and then substituting this $\bar{a}_{30}$ into the expression of v_2 results in $v_2 = \frac{1}{9}\,(\bar{b}_N + 6\,\bar{b}_D\,\bar{b}_{30})$ where

$$\bar{b}_D = a\,(8\,\bar{a}_{21} + 8\,a^2 - 8\,b^2 + 1) + 10\,\bar{b}_{21}\,(2\,b^2 - 2\,a^2\,b - 1 - 2\,\bar{a}_{21}). \tag{3.15}$$

There are two cases in solving the equation $v_2 = 0$. Since $b \neq 0$, so if $\bar{b}_D \neq 0$ one can uniquely determine $\bar{b}_{30}$, which is the generic case. When $\bar{b}_D = 0$, which is referred to as a special case, $\bar{b}_{30}$ (or b_{30}) is not used at this order. One may solve b_{21} from $\bar{b}_D = 0$ and then simplify v_2, from which a is determined as

$$a = \pm\left[\frac{16\,b^4 + 8\,b^3 a_{21} - 24\,b^2 a_{21}^2 - 14\,b^2 - 6\,a_{21} + 9}{20\,(2\,b^2 - 2\,b\,a_{21} - 1)}\right]^{1/2}. \tag{3.16}$$

Then, the expressions of the focus values, v_3, v_4 and v_5 can be greatly simplified. With a given above, v_3, v_4 and v_5 can be written as $v_3 = \frac{2\,a\,b^2\,F\,F_1}{625\,(2\,b^2 - 2\,b\,a_{21} - 1)^4}$, $v_4 = \frac{-\,2\,a\,b^2\,F\,F_2}{3515625\,(2\,b^2 - 2\,b\,a_{21} - 1)^7}$, $v_5 = \frac{a\,b^2\,F\,F_3}{31640625000\,(2\,b^2 - 2\,b\,a_{21} - 1)^{10}}$, where F is a common factor, and $F_i = F_i(b_{30}, a_{21}, b)$, $i = 1, 2, 3$, are i^{th}-degree polynomials of b_{30}. Since the common factors lead to centers, we must use F_i to solve the three equations: $v_3 = v_4 = v_5 = 0$. One can first explicitly solve b_{30} from $F_1 = 0$ and then simplify F_2 and F_3 to obtain two equations: $F_2(a_{21}, b) = 0$ and $F_3(a_{21}, b) = 0$, from which one obtains a polynomial equation: $F_4(b^2) = 0$ which in turn yields a total of four solutions for the special case:

$$\left.\begin{array}{llll} b = & 1.3798788398, & a_{21} = & 1.0897036998, & b_{30} = & 0.7093364483 \\ b = & -1.3798788398, & a_{21} = & -1.0897036998, & b_{30} = & -0.7093364483 \end{array}\right\}$$

$$a = \pm 0.0877426100, \quad b_{21} = \pm 0.1470131077, \quad a_{30} = \mp 0.0284721124. \qquad (3.17)$$

It is seen from Eq. (3.17) that all the four solutions satisfy $|b| > |a|$, indicating that the origin is a saddle point.

Then choosing any one of the four solutions, we can prove the existence of 12 small limit cycles [88]. A numerical simulation result with proper small perturbations is shown in Fig. 1, where 6 limit cycles are located in the neighborhood of each of the two fine focus points $(0, 1)$ and $(0, -1)$.

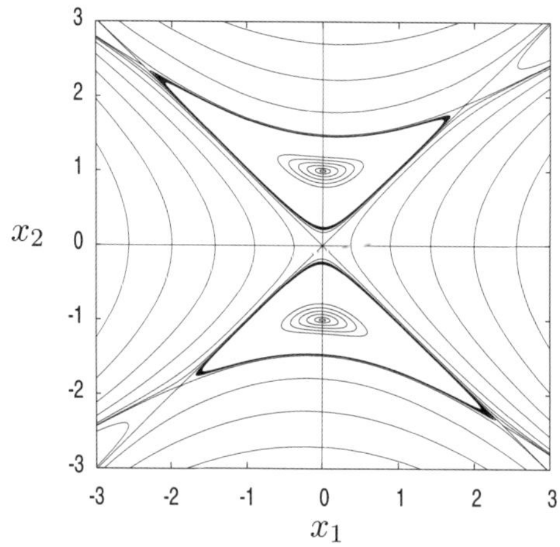

Figure 1 The phase portrait of the system (3.1) showing 12 limit cycles, 6 around each of the two fine focus points $(0,1)$ and $(0,-1)$, when the origin is a saddle point for $a = a_{12} = -b_{03} = -0.0883176873$, $b = -a_{03} = 1.3898788398$, $a_{30} = 0.0282840060$, $a_{21} = 1.1005496998$, $b_{30} = 0.7230834412$, $b_{21} = -0.1483513686$, $b_{12} = -1.0189112324$.

3.3 Existence of 12 limit cycles in the generic case. In this subsection, we turn to the generic case in which $\bar{b}_D \neq 0$ (see Eq. (3.15)). For this case, there are six free coefficients: a, b, $\bar{a}_{21}$, $\bar{b}_{21}$, $\bar{a}_{30}$ and $\bar{b}_{30}$, involved in the expressions of the focus values. Therefore, unlike the special case, it is now possible to have $v_i = 0$, $i = 1, 2, \ldots, 6$, but $v_7 \neq 0$. This indicates that it may be possible to have 7 small limit cycles around each of the focus points $(0, 1)$ and $(0, -1)$. We need to solve the four very large, coupled polynomial equations: $v_i = 0$, $i = 3, 4, 5, 6$. This causes very high complexity in computation.

First, letting $v_1 = 0$ yields the solution for $\bar{a}_{30}$, given by Eq. (3.14). Then, setting $v_2 = 0$ result in $\bar{b}_{30} = \bar{b}_{30}(a, b, \bar{a}_{21}, \bar{b}_{21}) = -\bar{b}_N/6/\bar{b}_D$. Next solving $v_3 = 0$ determines $\bar{a}_{21} = \bar{a}_{21}$. Then v_4, v_5 and v_6 are further simplified. However, these three polynomial equations are coupled and have to be solved simultaneously. Similarly, we can follow the procedure presented in the previous subsection to eliminate b to obtain two equations, and then another elimination of $\bar{b}_{21}$ from these two equations results in a single variable polynomial equation for a:

$$\begin{aligned} 0 = a\,(&112783323904064139036783273943 04a^{12} + 9439268357932268111521578554849280a^{10} \\ &+ 1595554914969941771013315201669920 0a^{8} + 7806733831958566794606224301855600a^{6} \\ &+ 2689670124233149942096312307701500a^{4} + 1317907196524296813182912146670625a^{2} \\ &+ 248225087542552407942139964497500). \end{aligned}$$

It is easy to see that the above equation has only one real solution $a = 0$. But if $a = 0$, then $\bar{b}_{21} = 0$, resulting in $\bar{a}_{21} = \bar{b}_{30} = \bar{a}_{30} = 0$, which can be shown to give a center. Thus, a must be non-zero. This implies that one cannot find possible parameter values such that $v_i = 0$, $i = 1, 2, \ldots, 6$. Hence, *14 small limit cycles are not possible* even for the generic case when the origin is a saddle point, and the maximal number of the small limit cycles for the generic case is still 12.

Similarly, we can also find all the solutions of the 12 limit cycles for the generic case, listed below: two solutions for the case when the origin is a saddle point:

$$\begin{aligned}
b &= \pm 15.7264394069\, a, \\
\bar{b}_{21} &= -1.1061229255\, a, \\
\bar{a}_{21} &= 0.7000000000 + 103.3880431509\, a^2, \\
\bar{b}_{30} &= \frac{0.2564102564\, a^2 (0.0089196607 + 0.0982810312\, a^2 + 0.2527227926\, a^4)}{0.0661528794 + 0.2704669566\, a^2}, \\
\bar{a}_{30} &= \mp(0.0806130156 - 17.7870588470\, a^2);
\end{aligned} \tag{3.18}$$

and two for the case when the origin is a node:

$$\begin{aligned}
b &= \pm 0.4765747114\, a, \\
\bar{b}_{21} &= 0.2033343806\, a, \\
\bar{a}_{21} &= 0.7000000000 + 1.0149654014\, a^2, \\
\bar{b}_{30} &= \frac{a^2 (0.0481488581 + 65.9546167690\, a^2 - 9379.2591506305\, a^4)}{0.0008286738 - 0.1076372236\, a^2}, \\
\bar{a}_{30} &= \pm(0.8202076319 + 2.4368685248\, a^2).
\end{aligned} \tag{3.19}$$

Further, for any of the above solutions, we can show the existence of 12 small limit cycles [87]). A numerical example is given for the case when the origin is a node, with $a = -0.7$. The phase portrait is depicted in Fig. 2, where the two boxes contain the 12 small limit cycles near the focus points $(0, \pm 1)$.

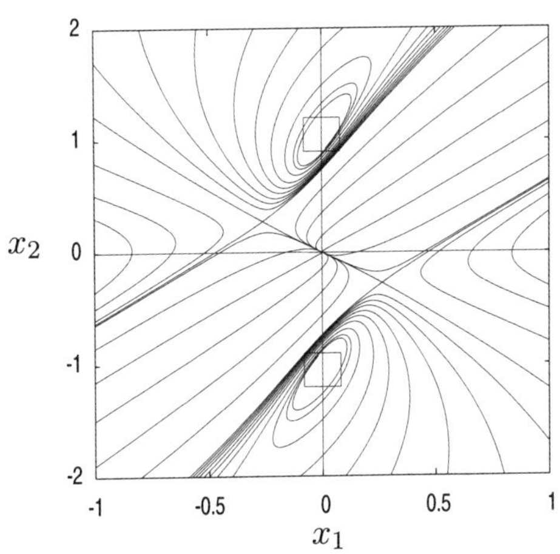

Figure 2 The phase portrait of the system (3.1) showing 12 limit cycles, 6 around each of the two fine focus points $(0, 1)$ and $(0, -1)$, when the origin is a node for $a = a_{12} = -b_{03} = -0.7$, $b = -a_{03} = -0.3336019980$, $a_{30} = 4.5658610164$, $a_{21} = -5.0539492766$, $b_{30} = -1.3447323014$, $b_{21} = -0.8333681006$, $b_{12} = -4.1028222711$.

3.4 Existence of small limit cycles in Z_2-equivariant vector fields. In the previous two subsections, we have discussed a special case and the generic case of a cubic order system, which exhibit 12 limit cycles. It has been noted that while the origin can only be a saddle point for the special case, it can be either a saddle point or a node for the generic case. It has been shown that for the two cases the maximal number of small limit cycles is 12. Recall in Eq. (3.3) that b_{12} has been chosen to make the frequency of the limit cycles equal 1. One may think that if let b_{12} be free, that is, the frequency can be chosen arbitrarily, then one might have 8 (in total 16 for the whole system) small limit cycles, since there are seven coefficients a, b, a_{21}, b_{21}, a_{30}, b_{30} and b_{12}, which can be chosen arbitrarily in the expressions of the focus values. This may give the possibility: $v_i = 0$, $i = 1, 2, \ldots, 7$, but $v_8 \neq 0$. In this subsetion, we shall prove that for any of the three cases, the maximal number of small limit cycles is 12.

To begin with, we use the real general Z_2-equivariant vector fields up to 3rd order [35, 87, 88]:

$$
\begin{aligned}
\dot{w}_1 = {} & (a_0 + a_3)\, w_1 - (b_0 - b_3)\, w_2 + (a_1 + a_4 + a_6 + a_8)\, w_1^3 \\
& - (b_1 - b_4 + 3\, b_6 - 3\, b_8)\, w_1^2\, w_2 + (a_1 + a_4 - 3\, a_6 - 3\, a_8)\, w_1\, w_2^2 \\
& - (b_1 - b_4 - b_6 + b_8)\, w_2^3\,, \\
\dot{w}_2 = {} & (b_0 + b_3)\, w_1 + (a_0 - a_3)\, w_2 + (b_1 + b_4 + b_6 + b_8)\, w_1^3 \\
& + (a_1 - a_4 + 3\, a_6 - 3\, a_8)\, x^2\, w_2 + (b_1 + b_4 - 3\, b_6 - 3\, b_8)\, w_1\, w_2^2 \\
& + (a_1 - a_4 - a_6 + a_8)\, w_2^3.
\end{aligned}
\tag{3.20}
$$

The two eigenvalues of the Jacobian of Eq. (3.20) evaluated at the origin $(w_1, w_2) = (0, 0)$ are $a_0 \pm \sqrt{a_3^2 + b_3^2 - b_0^2}$, indicating that the origin $(0, 0)$ is a saddle point or a node when $a_3^2 + b_3^2 - b_0^2 \geq 0$, or a focus point or a center if $a_3^2 + b_3^2 - b_0^2 < 0$. When $a_3^2 + b_3^2 - b_0^2 = 0$, the origin is either a node or a double zero singular point.

First, we consider $a_3^2 + b_3^2 - b_0^2 \geq 0$ under which the origin is a saddle point or a node. For this case, introduce a linear parametric transformation into Eq. (3.20), and rename the coefficients of the third order terms in the resulting equations as a_{30}, a_{21}, a_{12}, a_{03}, b_{30}, b_{21}, b_{12} and b_{03} to obtain

$$
\begin{aligned}
\dot{x} &= a\, x + b\, y + a_{30}\, x^3 + a_{21}\, x^2\, y + a_{12}\, x\, y^2 + a_{03}\, y^3, \\
\dot{y} &= b\, x + a\, y + b_{30}\, x^3 + b_{21}\, x^2\, y + b_{12}\, x\, y^2 + b_{03}\, y^3,
\end{aligned}
$$

where $a = a_0$ and $b = \pm\sqrt{a_3^2 + b_3^2 - b_0^2}$. This is exactly the system (3.1), considered in the previous two subsections. Note that system (3.20) has 12 parameters: a_0, a_1, a_3, a_4, a_6, a_8, b_0, b_1, b_3, b_4, b_6, b_8, while system (3.1) has only 10 parameters: a, b, a_{30}, a_{21}, a_{12}, a_{03}, b_{30}, b_{21}, b_{12}, b_{03}. This is because system (3.1) has put its linear terms, via the linear transformation, in the form of two independent parameters since the vector field of a planar system in the neighborhood of a fixed point is completely determined by two independent eigenvectors.

It is easy to observe from Eq. (3.1) that the system is invariant under the rotation about the origin by an angle π. To study the existence of small limit cycles which the system may exhibit, one needs fine focus points to generate degenerate Hopf bifurcations. For a vector field with Z_2-symmetry, naturally the best way is to have two symmetric focus points about the origin. Thus, if N small limit cycles are found in the neighborhood of one focus point, the whole system would have $2N$ limit cycles.

In addition, it has been noticed that the two eigenvectors of the saddle point (the origin) are $(1,1)$ and $(-1,1)$, so the two symmetric focus points can be assumed on the y-axis (or x-axis). Furthermore, by a proper scaling, one may suppose that the two focus points are located at $(0,\pm 1)$, leading to the conditions $a_{03} = -b$ and $b_{03} = -a$. Another condition $a_{12} = a$ comes from making the two focus point be Hopf type. These three conditions are exactly given in Eq (3.2).

Next, we shall show that the condition given in Eq. (3.3): $b_{12} = \frac{1+4a^2}{2b} - b$, which normalizes the frequency of limit cycles to be 1, is not necessary. In other words, if we assume the frequency can be selected arbitrarily, that is, b_{12} is free, then we can prove that this extra free coefficient does not increase possibility of having extra limit cycles. In general, one of the coefficients b_{12}, a and b (not necessarily b_{12}), can be chosen arbitrarily, which does not affect the results.

To prove the above statement, let ω be an arbitrary frequency of the limit cycles. Then, first introducing the following parameter scaling:

$$a \to \omega a, \quad b \to \omega b, \quad a_{30} \to \omega a_{30}, \quad a_{21} \to \omega a_{21}, \quad b_{30} \to \omega b_{30}, \quad b_{21} \to \omega b_{21}, \tag{3.21}$$

and then applying a transformation and a time scaling $\tau = \omega t$ finally gives the following new system:

$$\begin{aligned} \frac{du}{d\tau} &= v + 2\,\bar{a}_{21}\,u^2 + 4\,a\,u\,v - \tfrac{3}{2}\,v^2 + 4\,b\,\bar{a}_{30}\,u^3 - 2\,\bar{a}_{21}\,u^2 v - 2\,a\,u\,v^2 + \tfrac{1}{2}\,v^3, \\ \frac{dv}{d\tau} &= -u - 4\bar{b}_{21}u^2 + 2(2a^2-2b^2+1)uv - 8\bar{b}_{30}u^3 + 4\bar{b}_{21}u^2v - (2a^2-b^2+1)uv^2. \end{aligned} \tag{3.22}$$

which is exactly in the same form of Eq. (3.13), except that the time now is τ.

Therefore, based on the above discussion and the results obtained in the previous two subsections, we can conclude that for the case $a_3^2 + b_3^2 - b_0^2 \geq 0$ (i.e., the origin is either a saddle point or a node), the maximal number of small limit cycles that system (3.2) can exhibit is 12.

Similarly, we can discuss the case when $a_3^2 + b_3^2 - b_0^2 < 0$, and obtain an equation in the same form of (3.22). This indicates that the maximal number of small limit cycles for this case is also 12. Other degenerate cases can be similarly discussed. Summarizing the above results gives the following theorem.

Theorem 3.3 *The number of limit cycles that a cubic order system can have is at least* 12, *that is,* $H(3) \geq 12$.

To end this section, present a numerical example which exhibits 12 small limit cycles when the origin is also a focus point, with $a = -0.7$. The phase portrait of this example is shown in Fig. 3, where the areas near the focus points $(0,\pm 1)$ marked by the two boxes, contain the 12 small limit cycles.

4 Global problem – 80 limit cycles of a 9th-order system

For higher-order system (1.3), in order to obtain more limit cycles and various configuration patterns of their relative dispositions, we introduce an efficient method (see [39, 43, 38, 40, 41, 37]) to perturb symmetric Hamiltonian systems to have maximal number of centers, i.e., to study the weakened Hilbert's 16th problem for symmetric planar polynomial Hamiltonian systems, since bifurcation and symmetry are closely connected and symmetric systems play pivotal roles as a bifurcation point in all planar Hamiltonian systems. To investigate perturbed Hamiltonian systems, first we need to know the global behavior of the unperturbed polynomial systems, namely, to determine the global property of the families of real planar algebraic curves defined by the Hamiltonian functions. Then by using proper

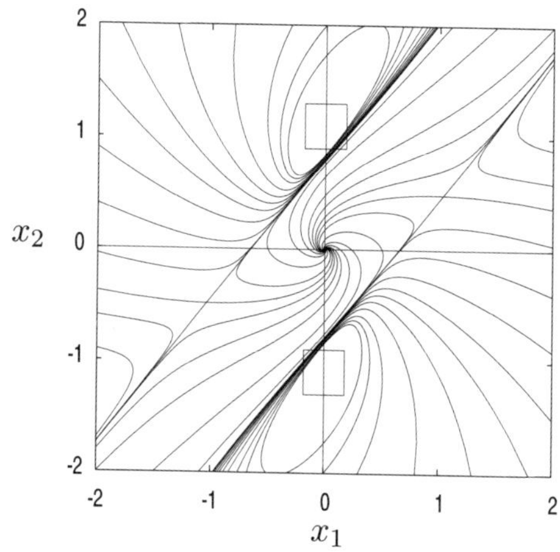

Figure 3 The phase portrait of the system (3.1) showing 12 limit cycles, 6 around each of the two fine focus points $(0, 1)$ and $(0, -1)$, when the origin is a focus point for $a = a_{12} = -b_{03} = -0.7$, $b = -a_{03} = -0.5889218635$, $a_{30} = 2.2161956860$, $a_{21} = -2.5682071892$, $b_{30} = -0.7072960219$, $b_{21} = 2.2961669830$, $b_{12} = -3.1019886923$.

perturbation techniques, we shall obtain the global information of bifurcations for the perturbed non-integrable systems. In this sense, we say that our study utilizes both the two parts of Hilbert's 16th problem.

On the basis of the method of detection function introduced by Li [39], we developed [75, 76, 77] a method of control parameters in order to obtain more limit cycles for a Z_{10}-equivariant perturbed polynomial Hamiltonian system of degree 9. With the aid of numerical simulations (using Maple) we show that there exist parameter groups such that our system has at least 80 limit cycles with three different configurations.

4.1 Z_q-equivariant planar vector fields. First, we present some basic concepts and results which will be used in the following subsections. Let G be a compact Lie group of transformations acting on R^n. A mapping $\Phi : R^n \to R^n$ is called G-equivariant if for all $g \in G$ and $x \in R^n$, $\Phi(gx) = g\,\Phi(x)$. A function $H : R^n \to R$ is called G-invariant function if for all $g \in G$ and $x \in R^n$, $H(gx) = H(x)$. If Φ is a G-equivariant mapping, the vector field $\dot{x} = \Phi(x)$ is called a G-equivariant vector field.

Let q be an integer. A group Z_q-equivariant vector fields is called a cyclic group of order q if it is generated by a planar counterclockwise rotation of the vector fields through $2\pi/q$ about the origin.

Introducing the transformation $z = x + iy$, $\bar{z} = x - iy$ into system (1.3) yields

$$\dot{z} = F(z, \bar{z}), \qquad \dot{\bar{z}} = \bar{F}(z, \bar{z}), \tag{4.1}$$

where $F(z, \bar{z}) = P(u, v) + i\,Q(u, v)$, $u = (z + \bar{z})/2$ and $v = (z - \bar{z})/(2\,i)$.

Theorem 4.1 [43] *A vector field defined by (4.1) is Z_q-equivariant, if and only if the function $F(z, \bar{z})$ has the following form:*

$$F(z, \bar{z}) = \sum_{l=1} g_l\,(|z|^2)\bar{z}^{lq-1} + \sum_{l=0} h_l\,(|z|^2) z^{lq+1}, \tag{4.2}$$

where $g_l(|z|^2)$ and $h_l(|z|^2)$ are polynomials with complex coefficients. In addition, Eq. (4.1) is a Hamiltonian system having Z_q-equivariance, if and only if Eq. (4.2) holds and

$$\frac{\partial F}{\partial z} + \frac{\partial \bar{F}}{\partial \bar{z}} \equiv 0. \tag{4.3}$$

Theorem 4.2 (Z_q–invariant function [35]) *A Z_q-invariant function $I(z, \bar{z})$ has the following form:*

$$I(z, \bar{z}) = \sum_{l=0} g_l(|z|^2) z^{lq} + \sum_{l=1} h_l(|z|^2) \bar{z}^{lq}. \tag{4.4}$$

One of the main questions in real algebraic geometry is: what schemes of the mutual arrangement (schemes or configurations) of ovals can be realized by the curves of a given degree? By using some Z_q-equivariant Hamiltonian systems, we can realize various configurations of ovals for planar algebraic curves of degree m.

4.2 The method of detection function: perturbations of Z_q-equivariant Hamiltonian systems. Consider the following perturbed Hamiltonian system:

$$\begin{aligned} \dot{x} &= \frac{\partial H}{\partial y} - \epsilon\, x(p(x,y) - \lambda), \\ \dot{y} &= -\frac{\partial H}{\partial x} - \epsilon\, y(p(x,y) - \lambda), \end{aligned} \tag{4.5}$$

where $H(x,y)$ is the Hamiltonian, $p(0,0) = 0$, $0 < \epsilon \ll 1$ and $\lambda \in R$. Suppose that the origin in the phase plane is a singular point of Eq. (4.5) and the following conditions hold:

(A1) The unperturbed system (4.5) ($\epsilon = 0$) is a Z_q-equivariant Hamiltonian vector field. There are families of curves $\{\Gamma^h\}$ defined by the Hamiltonian function $H(x,y) = h$ lying in periodic annuli which enclose at least one singular point. As the value of h changes toward the value defined by the singular points, Γ^h approaches a singular point or an inner boundary of the periodic annuli which consist of a heteroclinic (or homoclinic) loop.

(A2) Surrounding the periodic annuli, there exist heteroclinic (or homoclinic) loops which connect some hyperbolic saddle points at the values of h defined by these singular points.

(A3) The divergence $2\epsilon(\lambda - F(x,y)) \equiv 2\epsilon[\lambda - \frac{x}{2}\frac{\partial p}{\partial x} - \frac{y}{2}\frac{\partial p}{\partial y} - p(x,y)]$ of the perturbed vector field is a Z_q-invariant function.

Then, we can define

$$\lambda = \lambda(h) = \left(\iint_{D^h} F(x,y)\, dxdy\right) / \left(\iint_{D^h} dxdy\right) = \psi(h)/\phi(h), \tag{4.6}$$

which is called a detection function corresponding to the periodic family $\{\Gamma^h\}$. The graph of $\lambda = \lambda(h)$ in the plane (h, λ) is called a detection curve, where D^h is the area inside Γ^h.

Clearly, if $H(x,y) = h$ is a polynomial, then $\lambda(h)$ is a ratio between two Abelian integrals. In this case, $\lambda(h)$ is a differentiable function with respect to h. However, when the degree of $H(x,y)$ is greater than 4, classical mathematical analyses cannot be used to calculate $\lambda(h)$. In general, numerical techniques are applied to compute these Abelian integrals.

Based on the Poincaré-Pontrjagin-Andronov theorem on the global center bifurcations and Melnikov method (see [79, 90, 18, 48, 56, 80]), we have the following two theorems (as in Li *et al.* [39, 38, 43, 40, 41, 37]):

Theorem 4.3 (Bifurcation of limit cycles) *Suppose that the conditions (A1) and (A3) hold. For a given* $\lambda = \lambda_0$, *considering the set* S *of the intersection points of the straight line* $\lambda = \lambda_0$ *and the curve* $\lambda = \lambda(h)$ *in the* (h, λ)*-plane, we have the following conclusions:*

(i) *If* S *consists of exactly one point* (h_0, λ_0) *and* $\lambda'(h) > 0$ (< 0), *then there exists a stable (unstable) limit cycle of system (4.5) near* Γ^{h_0}.

(ii) *If* S *consists of two points* (h_0, λ_0) *and* $(\tilde{h}_0, \lambda_0)$ *having* $\tilde{h}_0 > h_0$ *and* $\lambda'(\tilde{h}_0) > 0$, $\lambda'(h_0) < 0$, *then there exist two limit cycles near* $\Gamma^{\tilde{h}_0}$ *and* Γ^{h_0}, *respectively, with the former stable and the latter unstable.*

(iii) *If* S *contains a point* (h_0, λ_0) *and* $\lambda'(h_0) = \lambda''(h_0) = \cdots = \lambda^{(k-1)}(h_0) = 0$ *but* $\lambda^{(k)}(h_0) \neq 0$, *then system (4.5) has at most* k *limit cycles near* Γ^{h_0}.

(iv) *If* S *is empty, then system (4.5) has no limit cycles.*

Theorem 4.4 (Bifurcation parameter created by a heteroclinic or homoclinic loop) *Suppose that the conditions (A1), (A2) and (A3) hold. Then for* $0 < \epsilon \ll 1$, *when* $\lambda = \lambda(h_2) + O(\epsilon)$, *system (4.5) has a heteroclinic (or homoclinic) loop with* Z_q*−equivariance at* $h = h_2$.

4.3 A Z_{10}-equivariant planar vector field. From the results given in section 2, it is easy to find that for $n = 9$, system (1.3) is Z_{10}-equivariant. Suppose it is Hamiltonian, then by changing the polar axis and time scale, we have the following 5-parameter system:

$$\begin{aligned} \dot{r} &= b\, r^9 \sin(10\theta) = R(r, \theta), \\ \dot{\theta} &= 1 + c\, r^2 + d\, r^4 + e\, r^6 + [\, a + b \cos(10\theta)\,]\, r^8 = \Theta(r, \theta), \end{aligned} \tag{4.7}$$

which has the Hamiltonian:

$$H(r, \theta) = -\tfrac{1}{2}\, r^2 - \tfrac{1}{4}\, c\, r^4 - \tfrac{1}{6}\, d\, r^6 - \tfrac{1}{8}\, e\, r^8 - \tfrac{1}{10}\, [\, a + b \cos(10\theta)\,]\, r^{10}. \tag{4.8}$$

Without loss of generality, suppose that $(r, \theta) = (1, 0)$ is a singular point of Eq. (4.7). Then, from $\Theta(r, 0) = 0$ and $\Theta(r, \pi/10) = 0$, we have $a+b+c+d+e+1 = 0$ and

$$\begin{aligned} f_1(r) &= (a+b) r^8 + e r^6 + d r^4 + c r^2 + 1 = 0, \\ f_2(r) &= (a-b) r^8 + e r^6 + d r^4 + c r^2 + 1 = 0. \end{aligned}$$

Further, suppose that the following parametric condition G holds: $a = 6.138618693$, $b = 0.02861869300$, $c = -6.684452552$, $d = 16.14239117$, $e = -16.62517600$, then system (4.7) has 81 finite singular points at $(0, 0)$ and $(z_1, 0)$, $(z_2, 0)$, $(z_3, 0)$ $(z_4, 0)$, $(z_5, \pi/10)$, $(z_6, \pi/10)$, $(z_7, \pi/10)$, $(z_8, \pi/10)$ and their Z_{10}-equivariant symmetric points. Here, $z_1 = 0.65$, $z_2 = 0.7$, $z_3 = 0.885$, $z_4 = 1$, $z_5 = 0.6389658748$, $z_6 = 0.7454834324$, $z_7 = 0.8086840806$, $z_8 = 1.050231017$. As h is varied, the level curves $H(r, \theta) = h$ of the Hamiltonian defined by Eq. (4.8) give rise to a family of tenth algebraic curves in the affine real plane (see Fig. 4).

Now the values of Hamiltonian function at critical points are calculated as follows:

$$\begin{aligned} h_1 &= H(z_1, 0) = -0.05793645019, & h_2 &= H(z_2, 0) = -0.05790831192, \\ h_3 &= H(z_3, 0) = -0.0588607574, & h_4 &= H(z_4, 0) = -0.0578621301, \\ h_5 &= H(z_5, \pi/10) = -0.05786629736, & h_6 &= H(z_6, \pi/10) = -0.05769015970, \\ h_7 &= H(z_7, \pi/10) = -0.05773717706, & h_8 &= H(z_8, \pi/10) = -0.0502778305, \end{aligned}$$

and

$$-\infty < h_3 < h_1 < h_2 < h_5 < h_4 < h_7 < h_6 < h_8 < 0.$$

As h increases from $-\infty$ to 0, the schemes of ovals of the nine algebraic curves are varied, with different structures, which yields the following result: In the affine real plane, the integral curves of the Hamiltonian system (4.7) can realize the following oval schemes: 1, 11, 21, $\frac{10}{1}$, $\frac{20}{1}$ and $C(1,1,1)$.

4.4 Bifurcation parameter values of Z_{10}-equivariant perturbed Hamiltonian systems. In this subsection we consider the following perturbed Z_{10}-equivariant vector field:

$$\begin{aligned} \dot{r} &= b\,r^9 \sin(10\theta) - \epsilon\, r\,(5\,s\,r^8 + 4\,u\,r^6 + 3\,v\,r^4 + 2\,w\,r^2 - \lambda), \\ \dot{\theta} &= 1 + c\,r^2 + d\,r^4 + e\,r^6 + [\,a + b\,\cos(10\theta)\,]\,r^8. \end{aligned} \tag{4.9}$$

Corresponding to Eq. (4.9), the function $F(x,y)$ in the divergence of vector field of the hypothesis (A3) (see Section 4.2) has the form:

$$F(r,\theta) = 5\,s\,r^8 + 4\,u\,r^6 + 3\,v\,r^4 + 2\,w\,r^2. \tag{4.10}$$

We continue to consider the parametric condition G, i.e., the unperturbed system (4.7) has the phase portrait as shown in Fig. 4.

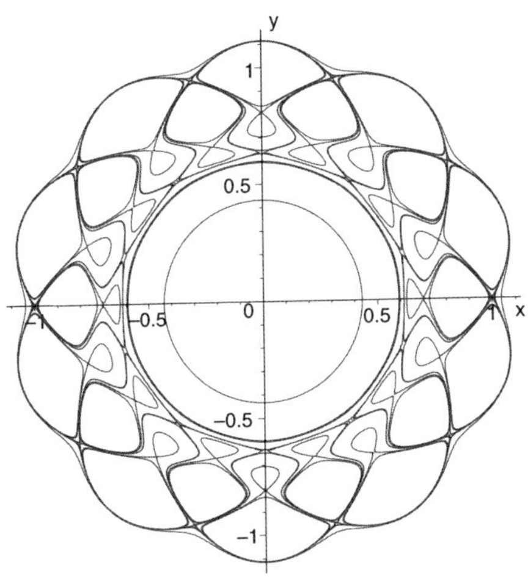

Figure 4 A Z_{10}-equivariant polynomial Hamiltonian vector field of degree 9.

We compute nine detection functions defined by Eq. (4.6), corresponding to the 9 types of periodic annuli $\{\Gamma_i^h\}$, $i = 0, 1, \ldots, 8$.

$$\begin{aligned} \lambda_i(h) &= \frac{\iint_{D_i^h} F(r,\theta)\, r\, dr\, d\theta}{\iint_{D_i^h},r\, dr\, d\theta} \\ &= \frac{\psi_i(h)}{\phi_i(h)} = \frac{1}{\phi_i(h)}\,[\,5\,s\,I_{i1}(h) + 4\,u\,I_{i2}(h) + 3\,v\,I_{i3}(h) + 2\,w\,I_{i4}(h)\,] \\ &= 5\,s\,J_{i1}(h) + 4\,u\,J_{i2}(h) + 3\,v\,J_{i3}(h) + 2\,w\,J_{i4}(h), \quad i = 0,\,1,\,\cdots,\,8, \end{aligned} \tag{4.11}$$

where $J_{ij}(h) = I_{ij}(h)/\phi_i(h)$, $j = 1, 2, 3, 4$ and $\phi_i(h) = \iint_{D_i^h} r\, dr\, d\theta$, in which D_i^h is the area inside Γ_i^h, and $I_{i1}(h) = \iint_{D_i^h} r^9\, dr\, d\theta$, $I_{i2}(h) = \iint_{D_i^h} r^7\, dr\, d\theta$, $I_{i3}(h) = \iint_{D_i^h} r^5\, dr\, d\theta$, $I_{i4}(h) = \iint_{D_i^h} r^3\, dr\, d\theta$. For the given parameter group G, the functions $J_{ij}(h)$ can be numerically calculated to a given degree of accuracy. In this paper, the accuracy is given up to 10 decimal points.

We can follow the approach described in [77] to find the parameter values for Hopf bifurcations, the bifurcations of heteroclinic or homoclinic loops, and bifurcation direction detections of heteroclinic and homoclinic loops. Then we determine the perturbed parameter group $PG : (s, u, v, w)$ such that our nine detection functions $\lambda_i(h)$, $i = 0, 1, \cdots, 8$, can create extra limit cycles. In other words, we can choose parameters to "control" the number and configuration of limit cycles created by Eq. (4.9). To this end, three limit cycle configurations have been obtained with different control groups.

Case 1. Suppose that the following conditions hold:

C1: $\lambda_1(h_2) - \lambda_2(h_2) = 0,$
C2: $\lambda_6(h_7) - \lambda_2(h_1) = 0,$
C3: $\lambda_3(h_5) - \lambda_7(h_7) = 0,$

which imply that $s = s$, $u = -3.568366338s$, $v = 4.880734178s$, $w = -3.201327275s$. As an example, taking $s = 2.8$, then

$$PG_1 = (s, u, v, w) = (2.8, \ -9.991425746, \ 13.66605570, \ -8.963716370),$$

and we obtain the bifurcation values of $\lambda_i(h)$ and the values of direction detections. Under the parameter conditions of G and PG_1, system (4.9) has the graphs of detection curves as shown in Fig. 5(a) and a zoom-in is depicted in Fig. 5(b). It is seen from Fig. 5(a) that when

$$\tilde{\lambda} \in (\lambda_7(h_6), \lambda_7(h_7)) = (-2.82477103, \ -2.82469595), \tag{4.12}$$

in the (h, λ)-plane the straight line $\lambda = \tilde{\lambda}$ intersects the curves $\lambda = \lambda_2(h)$, $\lambda_3(h)$, $\lambda_5(h)$ and $\lambda = \lambda_6(h)$ at least once, and intersects the curves $\lambda = \lambda_1(h)$ and $\lambda = \lambda_7(h)$ at least twice, which makes the total of 8 intersections in one of the ten equivariant sections. According to Theorem 4.4, there exist 8 limit cycles in one section. Since system (4.9) is Z_{10}-equivariance, it has at least 80 limit cycles with configuration shown in Fig. 6.

Case 2. If the control conditions given in Case 1 are changed to

C1: $\lambda_1(h_2) - \lambda_5(h_7) = 0,$
C2: $\lambda_6(h_7) - \lambda_1(h_3 + 0.00065) = 0,$
C3: $\lambda_2(h_1 + 0.00001) - \lambda_1(h_1 + 0.000005) = 0,$

then, another parameter group PG_2 is obtained as:

$$PG_2 = (s, u, v, w) = (2.803157747, \ -10, \ 13.67228371, \ -8.962712462),$$

under which the bifurcation values of $\lambda_i(h)$ and the values of bifurcation direction detections can be found and the graphs of detection curves shown in Fig. 7(a) and a zoom-in is given in Fig. 7(b). It is observed that when $\tilde{\lambda} \in (\lambda_6(h_6), \ \lambda_6(h_7)) = (-2.82243550, -2.82235555)$, in the (h, λ)-plane the straight line $\lambda = \tilde{\lambda}$ intersects the curves $\lambda = \lambda_1(h)$, $\lambda_2(h)$, $\lambda_5(h)$ and $\lambda = \lambda_6(h)$ at least once, and intersects the curves $\lambda = \lambda_3(h)$ and $\lambda = \lambda_7(h)$ at least twice, which makes the total of 8 intersections in one of the ten equivariant sections. Therefore, it comes out 8 limit cycles in one section and 80 for the whole plane, as shown in Fig 8.

Case 3. If the control conditions are changed to

C1: $\lambda_1(h_3) - \lambda_4(h_7) = 0,$
C2: $\lambda_5(h_7 + 0.000037) - \lambda_4(h_7) = 0,$
C3: $\lambda_2(h_1 + 0.00001) - \lambda_1(h_1 + 0.000005) = 0,$

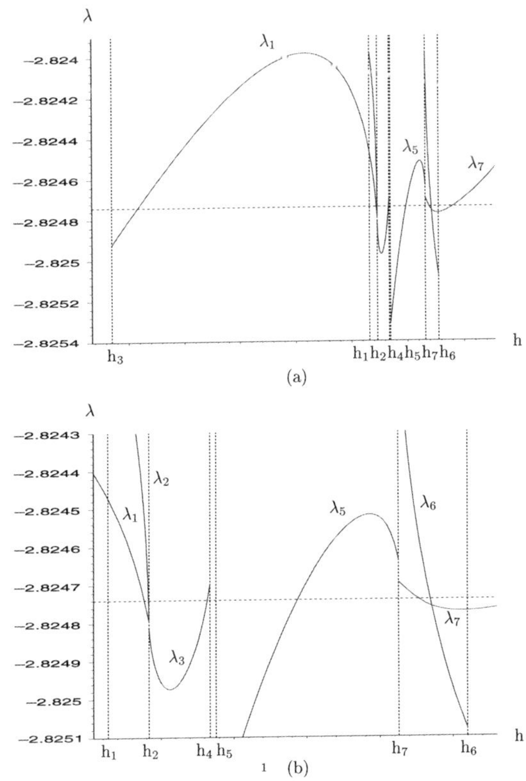

Figure 5 (a) Graphs of detection curves of Eq. (4.9) for parameters G and GP_1 with $h_3 = -0.0588607574$, $h_1 = -0.05793645019$, $h_2 = -0.05790831192$, $h_4 = -0.0578621301$, $h_5 = -0.05786629736$, $h_7 = -0.05773717706$, $h_6 = -0.05769015970$; and (b) a Zoom-in of part (a).

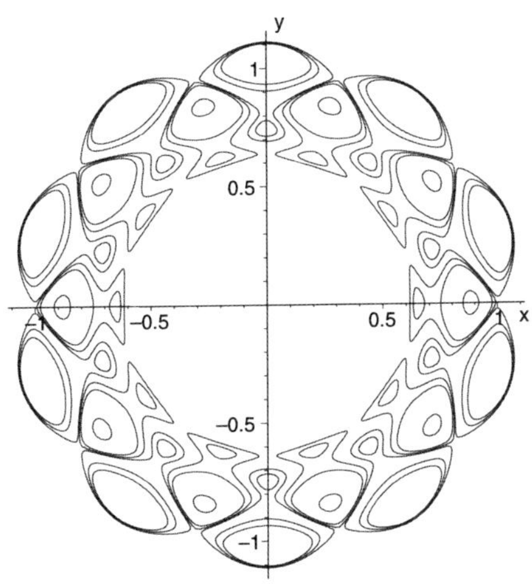

Figure 6 Existence of 80 limit cycles in system (4.9) with GP_1.

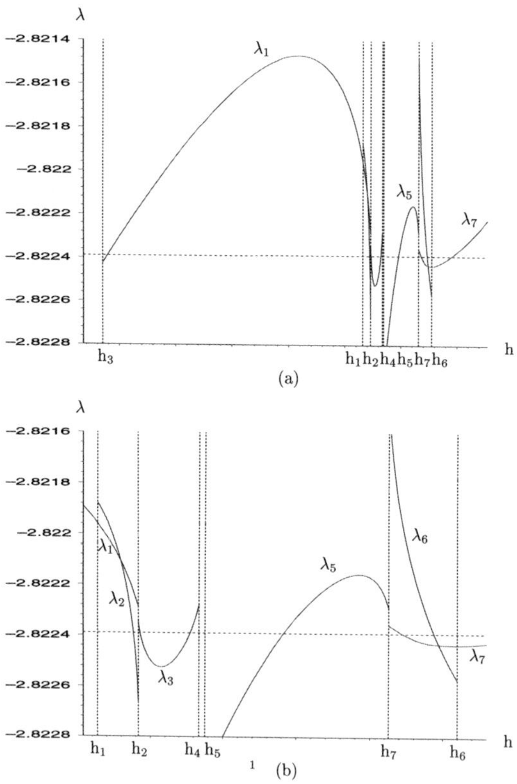

Figure 7 (a) Graphs of detection curves of Eq. (4.9) for parameters G and GP_2 with $h_3 = -0.0588607574$, $h_1 = -0.05793645019$, $h_2 = -0.05790831192$, $h_4 = -0.0578621301$, $h_5 = -0.05786629736$, $h_7 = -0.05773717706$, $h_6 = -0.05769015970$; and (b) a Zoom-in of part (a).

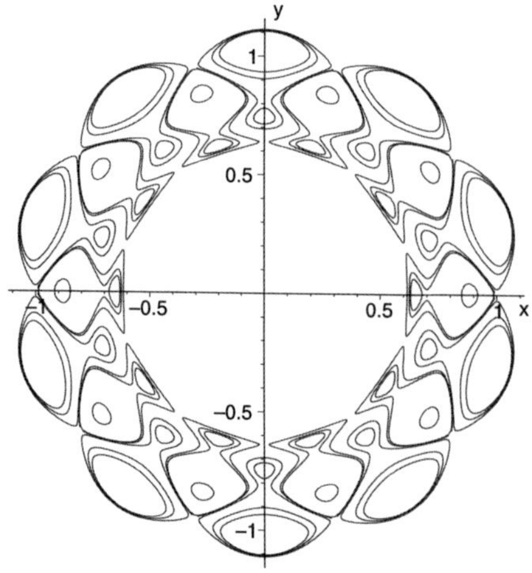

Figure 8 Existence of 80 limit cycles in system (4.9) with GP_2.

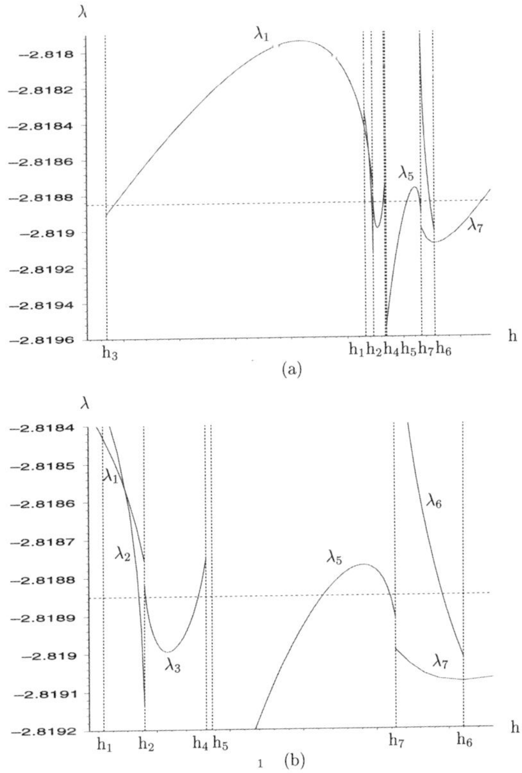

Figure 9 (a) Graphs of detection curves of Eq. (4.9) for parameters G and GP_3 with $h_3 = -0.0588607574$, $h_1 = -0.05793645019$, $h_2 = -0.05790831192$, $h_4 = -0.0578621301$, $h_5 = -0.05786629736$, $h_7 = -0.05773717706$, $h_6 = -0.05769015970$; and (b) a Zoom-in of part (a).

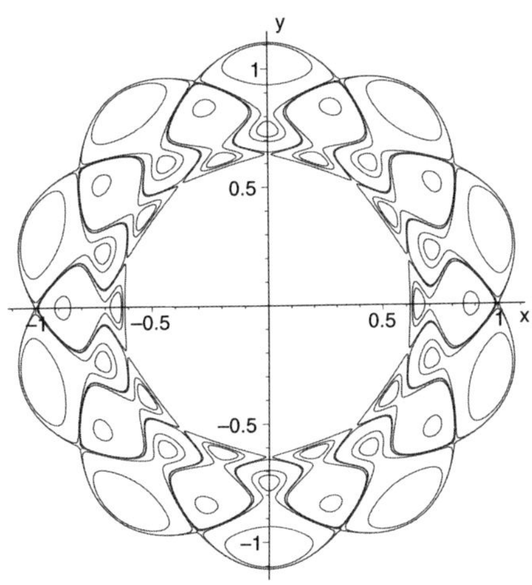

Figure 10 Existence of 80 limit cycles in system (4.9) with GP_3.

then, another parameter group PG_3 is obtained as:

$$PG_3 = (\,s, u, v, w) = (2.8, \,-9.988759290, \; 13.65655815, \,-8.951989483\,),$$

for which the detection function curves shown in Fig. 9(a) and see a zoom-in in Fig. 9(b). When $\tilde{\lambda} \in (\lambda_5(h_7),\ \lambda_3(h_2)) = (-2.82228573,\ -2.82234534)$, in the (h, λ)-plane the straight line $\lambda = \tilde{\lambda}$ intersects the curves $\lambda = \lambda_1(h)$, $\lambda_2(h)$, $\lambda_6(h)$ and $\lambda = \lambda_7(h)$ at least once, and intersects the curves $\lambda = \lambda_3(h)$ and $\lambda = \lambda_5(h)$ at least twice, which makes a total of 8 intersections in one of the ten equivariant sections. Therefore, it again gives 8 limit cycles in one section and 80 for the whole plane, as shown in Fig. 10.

Consequently, the main results of this section is proved, summarized in the following theorem.

Theorem 4.5 *For the unperturbed and perturbed parameter groups G and PG, and small $\epsilon > 0$, when $\lambda = \tilde{\lambda}$ satisfies (4.12), system (4.9) has at least* 80 *limit cycles, i.e., $H(9) \geq 80$.*

5 Conclusion

A detailed study has been carried out on the bifurcation of limit cycles for planar polynomial systems. After a literature review given on the second part of Hilbert's 16th problem, the main attention has been given to the 3rd-degree and 9th-degree polynomial systems. It has been shown using normal form theory that general cubic order, Z_2-equivariant vector fields can have three typical cases with 12 limit cycles, implying that $H(3) \geq 12$. For the vector fields of degree 9 with Z_{10}-symmetry, global bifurcation analysis with the detection function method is employed to show that such a system can have at least 80 limit cycles, i.e., $H(9) \geq 9^2 - 1$. Very recently, various examples have been constructed to shown that $H(4) \geq 4^2 - 1$ [89], $H(5) \geq 5^2 - 1$ [36], $H(6) \geq 6^2 - 1$ [75], $H(7) \geq 7^2$ [42], and $H(11) \geq 11^2$ [76].

References

[1] Anosov, D. V., Arnold, V. I. (Eds.) *Dynamical systems I.* Encyclopaedia Math. Sci., Springer-Verlag, Berlin, 1988

[2] Arnold, V. I. *Loss of stability of self-oscillations close to resonance and versal deformations of equivariant vector fields.* Funct. Anal. Appl. **11** (1977), 85–92.

[3] Arnold, V. I., Afraimovich, V. S., Ilyashenko, Yu. S. and Shilnikov, L. P. *Bifurcation theory, dynamical systems,* vol. 5, Springer-Verlag, NY, 1994.

[4] Arnold, V. I. and Oleinik, O. A. *The topology of real algebraic varieties.* Moscow Univ. Math. Bull. **34** (1979), 7–17.

[5] Bamón, R. *Quadratic vector fields in the plane have a finite number of limit cycles.* Publ. Math. I.H.E.S. **64** (1986), 111–142.

[6] Bautin, N. [1952] *On the number of limit cycles which appear with a variation of coefficients from an equilibrium position of focus or center type.* Mat. Sb.(N.S.) **30** (1954), 181–196 (Russian); Transl. Amer. Math. Soc. **100**, 397–413.

[7] Bendixson, I. *Sur les courbes définies par les équations différentielles.* Acta Math. **24** (1901), 1–88.

[8] Chen, L. S. and Wang, M. S. *The relative position, and the number, of limit cycles of a quadratic differential system.* Acta. Math. Sinica **22** (1979), 751–758.

[9] Chicone, C. and Schafer, S. *Separatrix and limit cycles of quadratic systems and Dulac's theorem.* Trans. Amer. Math. Soc. **278** (1983), 585–612.

[10] Christopher, C. J. and Lloyd, N. G. *Polynomial systems: lower bound for the Hilbert numbers.* Proc. Royal. Soc. London Ser. A **450** (1995), 219–214.

[11] Coleman, C. S. *Hilbert's 16th problem: How many cycles?* Differential equation models, modules in applied mathematics (Springer-Verlag, NY) **1** (1983), 279–297.

[12] Dulac, H. *Sur les cycles limite.* Bull. Soc. Math. France **51** (1923), 45–188.

[13] Dumortier, F., Roussarie, R. and Rousseau, C. *Hilbert's 16th problem for quadratic vector field.* J. Diff. Eqns. **110** (1994), 86–133.

[14] Dumortier, F., Roussarie, R. and Rousseau, C. *Elementary graphics of cyclicity 1 and 2.* Nonlinearity **7** (1994), 1001–1043.

[15] Dumortier, F., Morsalani, M. and Rousseau, C. *Hilbert's 16th problem for quadratic vector fields and cyclicity of elementary graphics.* Nonlinearity. **9** (1996), 1209–1261.

[16] Écalle, J. *Introduction aux fonctions analysables et preuve constructive de la conjecture de Dulac*, (French) Hermann, Paris, 1992.

[17] Gavrilov, L. and Iliev, I. *Second-order analysis in polynomially perturbed reversible quadratic Hamiltonian systems.* Ergod. Th. & Dynam. Sys. **20** (2000), 1671–1686.

[18] Guckenheimer, J. and Holmes, P. *Nonlinear Oscillations, Dynamical Systems, and Bifurcations of vector Fields (2nd Ed.)*, Springer-Verlag, New York, 1992.

[19] Gudkov, D. A. *The topology of real projective algebraic varieties.* Russ. Math. Surv. **29** (1974), 3–79.

[20] Han, M., Lin, Y. and Yu, P. *A study on the existence of limit cycles of a planar system with 3rd-degree polynomials.* Int. J. Bifurcation & Chaos **14** (2004), 41–60.

[21] Harnack, A. *Ueber die Vieltheiligkeit der ebenen algebraischen Curven.* Math. Annalen **10** (1876), 189–198.

[22] Hilbert, D. *Mathematical problems.* (M. Newton, Transl.) Bull. Amer. Math. Soc. **8** (1902), 437–479.

[23] Hilbert, D. *Mathematical problems.* Proc. Sympos. Pure Math. **28** (1976), 1–34.

[24] http://www.brainyencyclopedia.com.

[25] Iliev, I. *High-order Melnikov functions for degenerate cubic Hamiltonians.* Adv. Diff. Eqns. **1** (1996), 689–708.

[26] Ilyashenko, Yu. S. *The origin of limit cycles under perturbation of the equation* $dw/dz = -R_z/F_w$, *where* $R(z,w)$ *is a polynomial.* Math. USSR-SB. **7** (1969), 353–364.

[27] Ilyashenko, Yu. S. *Centennial history of Hilbert's 16th problem.* Amer. Math. Soc. **39** (2002), 301–354.

[28] Ilyashenko, Yu. S. [1982] *Singular points and limit cycles of differential equations in the real and the complex planes.* (Preprint) Pushchino: Sci. Res. Comput. Cent., Acad. Schi. USSR. 39pp (in Russian).

[29] Ilyashenko, Yu. S. *Dulac's memoir "On limit cycles" and related questions of the local theory of differential equations.* (Russian) Uspekhi Mat. Nauk **40** (1985), 41–78.

[30] Ilyashenko, Yu. S. and Yakovenko, S. [1991] *Finitely smooth normal forms of local families of diffeomorphismes and vector fields.* Russ. Math. Surv. **46** (1991), 3–19.

[31] Ilyashenko, Yu. S. and Yakovenko, S. *Counting real zeros of analytic functions satisfying linear ordinary differential equations.* J. Diff. Eqns. **126** (1996), 613–650.

[32] Kaloshin, V. Yu. *The Hilbert-Arnold problem and estimate of the cyclicity of polycycles on the plane and in space.* Funct. Anal. Appl. **35** (2001), 146–147.

[33] Khovanskii, A. *Real analytic manifolds with the property of finiteness, and complex Abelian integrals* (Russian). Funktsional. Anal. i Prilozhen. **18** (1984), 40–50.

[34] Korchagin, A. B. *Smoothing of 6-fold singular points and constructions of 9th degree M-curves.* Amer. Math. Soc. Trans. **173** (1996), 141–155.

[35] Li, J. B *Hilbert's 16th problem and Bifurcations of Planar polynomial vector fields.* Int. J. Bifurcations & Chaos **13** (2003), 47–106.

[36] Li, J. B., Chan, H. S. Y. and Chung, K. W. *Bifurcations of limit cycles in a* Z_6*-equivariant planar vector field of degree 5.* Science in China (Ser. A) **45** (2002), 817–826.

[37] Li, J. B., Chan, H. S. Y. and Chung, K. W. *Some lower bounds for* $H(n)$ *in Hilbert's 16th problem.* Qualitative Theory of Dynamical Systems **3** (2002), 345–360.

[38] Li, J. B. and Huang, Q. *Bifurcations of limit cycles forming compound eyes in the cubic system.* Chinese Ann. of Math. **8B** (1987), 391–403.

[39] Li, J. B. and Li, C. F. *Planar cubic Hamiltonian systems and distribution of limit cycles of* (E_3). Acta. Math. Sinica **28** (1985), 509–521.

[40] Li, J. B. and Liu, Z. R. *Bifurcation set and limit cycles forming compound eyes in a perturbed Hamiltonian system.* Publications Mathmatiques **35** (1991), 487–506 (in Spain).

[41] Li, J. B. and Liu, Z. R. *Bifurcation set and compound eyes in a perturbed cubic Hamiltonian system.* In Ordinary and Delay Differential Equations, π Pitman Research Notes in Math. Series vol. 272, Longman, England, 1992, pp. 116–128.
[42] Li, J. and Zhang, M. *Bifurcations of limit cycles in a Z_8-equivariant planar vector field of degree 7.* J. Diff. Equs. & Dyn. Sys. **16** (2004), 1123–1139.
[43] Li, J. B and Zhao, X. H. *Rotation symmetry groups of planar Hamiltonian systems.* Ann. Diff. Eqns. **5** (1989), 25–33.
[44] Liénard, A. *Etude des oscillations entretenues Rev.* Generale de l'Electricite **23** (1928), 901–912.
[45] Lins, Neto A. *On the number of solutions of the equation $dx/dt = \sum_{j=1}^{n} a_j(t)x^j, 0 \leq t \leq 1$, for which $x(0) = x(1)$.* Invent. Math. **59** (1980), 67–76.
[46] Lins, Neto A., de Melo, W. and Pugh C. *On Lienard's equation.* Lecture Notes in Math. vol. 597, Springer Berlin, 1977, pp. 335–357.
[47] Lloyd, N. G. *Limit cycles of polynomial systems. In New Directions in Dynamical Systems.* (Ed. T. Bedford and J. Swift) London Math. Soc. Lecture Notes **40** (1988), 192–234.
[48] Luo, D. J., Wang, X., Zhu, D. M. and Han, M. *Bifurcation Theory and Methods of Dynamical Systems*, World Scientific, Singapore, 1997.
[49] Melnikov, V. K. *On the stability of the center for time periodic perturbations.* Trans. Moscow Math. Soc. **12** (1963), 1–57.
[50] Murphy, G. M. *Ordinary differential equations and their solution*, Princeton, NJ: Van Nostrand, 1960.
[51] Novikov, D. and Yakovernko, S. *Simple exponential estimate for the number of real zeros of complete abelian integrals.* Annales de l'institut Fourier **45**(4) (1995), 897–927.
[52] Oleinik, O. A. *On Hilbert's 16th problem*, in Hilbert's Problems, Nauka, Moscow, 1969, pp. 182–195.
[53] Orekov, S. Yu. [1999] *Link theory and oval arrangements of real algebraic curves.* Topology **38** (1999), 779–810.
[54] Otrokov, N. T. *On the number of limit cycles of a differential equation in a neighborhood of a singular point.* Mat. Sb. **34** (1954), 127–144 (in Russian).
[55] Panov, A. *Variety of Poincaré mappings for cubic equations with variable coefficients.* (Russian). Funktsional. Anal. i Prilozhen. **33** (1999), 84–88.
[56] Perko, L. M *Differential equations and dynamical systems*, Springer-Verlag, New York, 2000.
[57] Petrovskii, I. and Landis, E. *On the number of limit cycles of the equation $dy/dx = P(x,y)/Q(x,y)$, where P and Q are polynomials of 2nd degree.* (in Russian) Mat. Sb. N.S. **37** (1955), 209–250.
[58] Petrovskii, I. and Landis, E. *On the number of limit cycles of the equation $dy/dx = P(x,y)/Q(x,y)$, where P and Q are polynomials.* (in Russian) Mat. Sb. N.S. **85** (1957), 149–168.
[59] Poincaré, H. *Les Methodes Nouvelles de la Mecanique Celeste*, 1892-1899.
[60] Poincaré, H. *Mémoire sur les courbes définies par une équation différentielle, I, II.* J. Math. Pures Appl. **7** (1881), 375–422; **8** (1882), 251–296; *Sur les courbes définies par les équations différentielles, III, IV.* ibid. **1** (1885), 167–224; **2** (1886), 155–217.
[61] Polotovskii, G. M. *On the Classification of Decomposing Planar Algebraic Curves*, Lecture Notes in Mathematics vol. 1524, Springer, 1992, pp. 52–74.
[62] Pontryagin, L. S. *On dynamical systems close to Hamiltonian systems.* ZH. EKsp. Teor. Fiz. **4** (1934), 234–238 (in Russian).
[63] Rohn, K. *Flchen vierter Ordnung.* Preisschriften der Frstlich Jablonowskischen Gesellschaft, Leipzig, 1886.
[64] Roussarie, R. *A note on finite cyclicity and Hilbert's 16th problem*, in Dynamical Systems, Valparaiso 1986 (Eds. R. Bamom *et al.*) Lecture Notes in Mathematics, Springer-Verlag, NY **1331** (1988), 161–168.
[65] Roussarie, R. *Bifurcation of planar vector fields and Hilbert's 16th problem.* Progress in Mathematics vol. 164, Birkauser Verlag, Basel, 1998.
[66] Rousseau, C. *Bifurcation methods in polynomial systems*, in Bifurcations and Periodic Orbits of Vector Fields (Ed. D. Schlomiuk) NATO ASI Series C, Kluwer Academic, London **408** (1993), 383–428.
[67] Rousseau, C. and Schlomiuk, D. *Cubic vector fields symmetric with respect to a center.* J. Diff. Eqns. **123** (1995), 388–436

[68] Schlomiuk, D. *Algebraic and geometric aspects of the theory of polynomial vector fields*, in Bifurcation and Periodic Orbits of Vector Fields, NATO ASI Series C (Ed. D. Schlomiuk) **408** (1993), 429–467.
[69] Shahshahani, S. *Periodic solutions of polynomial first order differential equations.* Nonlinear Anal. **5** (1980), 157–165.
[70] Shi, S. *A concrete example of the existence of four limit cycles for plane quadratic systems.* Sci. Sinica **23** (1980), 153–158.
[71] Shustin, E. *Topology of Real Plane Algebraic Curves*, Lecture Notes in Mathematics vol. 1524, Springer, 1992, pp. 97–109.
[72] Smale, S. *Mathematical problems for the next century.* The Mathematical Intelligencer **20** (1998), 7–15.
[73] Varchenko, A. *Estimation of the number of zeros of an Abelian integral depending on a parameter, and limit cycles* (Russian). Funktsional. Anal. i Prilozhen. **18** (1984), 14–25.
[74] Viro, O. Ya. *Progress in the topology of real algebraic varieties over the last six years.* Russ. Math. Surv. **41** (1986), 55–82.
[75] Wang, S. and Yu, P. *Bifurcation of limit cycles in a quintic Hamiltonian system under sixth-order perturbation.* Chaos, Solitons & Fractals **26**(5) (2005), 1317–1335.
[76] Wang, S. and Yu, P. *Existence of 121 limit cycles in a perturbed planar polynomial Hamiltonian vector field of degree 11.* Chaos, Solitons & Fractals (2006) (in press).
[77] Wang, S., Yu, P. and Li, J. *Bifurcation of limit cycles in a $Z_1 0$-equivariant vector fields of degree 9.* Int. J. Bifurcations & Chaos **16**(7) (2006).
[78] Wilson G. [1978] Hilbert's 16th problem. *Topology.* **17**(1), 53-73.
[79] Ye, Y. Q. *Theory of Limit Cycles.* (Transl. Math. Monographs 66) Amer. Math. Soc., Providence, RI, 1986.
[80] Ye, Y. Q. *Qualitative theory of polynomial differential systems* (Modern Mathematics Series), Shanghai Scientific and Technical Publishers, Shanghai, 1995 (in Chinese).
[81] Ye, Y. Q. *On Hilbert's 16th problem in the case $n = 2$.* Ann. Diff. Eqns. **12** (1996), 110–116.
[82] Yu, P. *Computation of normal forms via a perturbation technique.* J. Sound & Vib. **211** (1998), 19–38.
[83] Yu, P. *Limit cycles in 3rd-order planar system.* Int. Congress of Mathematicians, Beijing, China, August 20-28, 2002.
[84] Yu, P. *Twelve limit cycles in a cubic case of the 16th Hilbert problem.* Workshop on Bifurcation Theory and Spatio-Temporal Pattern Formation in PDE, Toronto, Canada, December 11-13, 2003.
[85] Yu, P. *Study of limit cycles in planar systems.* International Workshop on Bifurcation Theory and Applications, Shanghai, China, May 23-26, 2004.
[86] Yu, P. and Han, M. *Twelve limit cycles in a 3rd-order planar system with Z_2 symmetry.* Communication on Applied and Pure Analysis **3** (2004), 515–526.
[87] Yu, P. and Han, M. *Twelve limit cycles in a cubic case of the 16th Hilbert problem.* Int. J. Bifurcations & Chaos **15**(7) (2005), 2191–2205.
[88] Yu, P. and Han, M. *Small limit cycles bifurcating from fine focus points in cubic order Z_2-equivariant vector fields.* Chaos, Solitons and Fractals **24** (2005), 329–348.
[89] Zhang, T., Han, M., Zang, H. and Meng, X. *Bifurcation of limit cycles for a cubic Hamiltonian system under quartic perturbations.* Chaos, Solitons and Fractals **22** (2004), 1127–1138.
[90] Zhang, Z. F., Ding, T. R., Huang, W. Z. and Dong, Z. X. *Qualitative theory of differential equations* (Transl.Math. Monographs 101), Amer. Math. Soc., Providence, RI, 1992.
[91] Zhang, T., Zang, H. and Han, M. *Bifurcations of limit cycles in a cubic system.* Chaos, Solitons and Fractals **20** (2004), 629–638.
[92] Zwillinger, D. [1997] *Handbook of differential equations* (3rd ed.) Boston, MA: Academic Press, p. 120, 1997.